ROCKS

ROCKS

A Guide to the Stones Around Us and the Stories They Tell

Vojta Hybl

CONTENTS

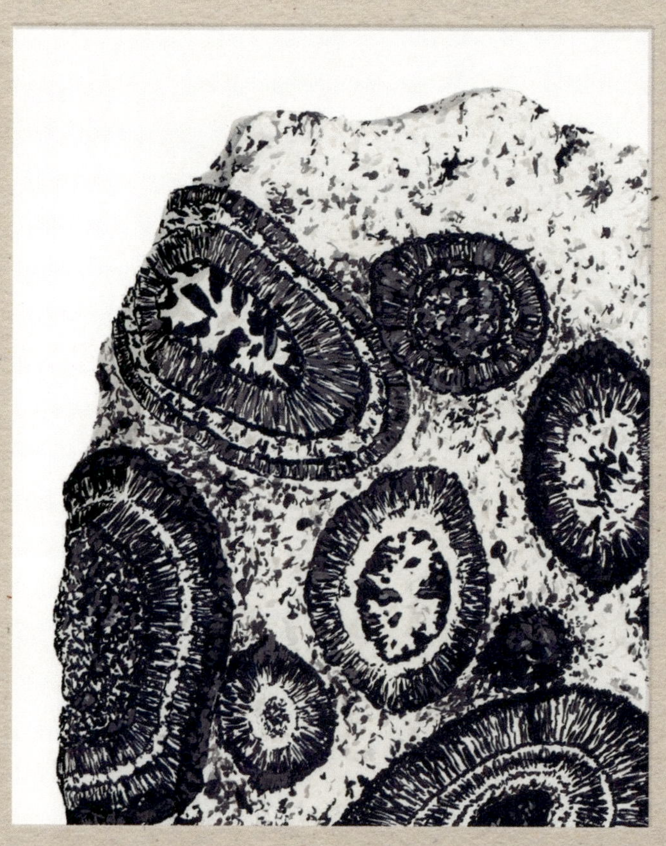

INTRODUCTION

Being able to understand rocks is a powerful thing. It offers us a different perspective on our world and helps us form a deeper, more meaningful relationship with the Earth. But let's not beat around the bush: a lot of rocks are hard to identify, which makes it difficult to appreciate the geological stories that they represent. Things are easier with most plants, animals and fungi – one species looks more or less the same no matter where you encounter it. However, the same rock type can look very different, not only in two distant places, but even within the same outcrop. And then there are rocks that appear to be the same but are actually not related at all. As a result, most of us don't give the lithosphere (that's the outer rocky layer of the Earth) a second glance; it becomes the non-living backdrop to everything else. We become rock-blind despite being surrounded by rocks.

It also doesn't help that geology has an image problem. Despite all the timely research that is being done in the areas of renewable energy, climate change and natural hazards (to name just a few), geology has for far too long been synonymous with irresponsible mining, damage to the environment and chasing profit at all costs. It is no surprise, though. As a scientific discipline, it developed from colonialism and the need to extract more and more natural resources (alongside the nobler intention to study the evolution of the Earth). Many of the problematic ways in which we think about the Earth are rooted in colonialism. European geologists – mostly white men – went all over the world to find resources for the benefit of their home countries, in the process using and erasing traditional Indigenous knowledge (as well as the contributions of women). Rocks were identified, classified and mapped to be taken.

Western thinking has taught us that rocks are just objects. However, geology offers so much more than that. *Gê* is the Ancient Greek word for 'earth', while *logía* denotes not only 'an area of study', but also 'words' or 'narratives'. The story of the Earth is quite literally written in rocks. Geological knowledge is one way we can understand it.

We can't change history, but we can change how we see the Earth today. I'd like to invite you to think beyond the human-centric point of view. Let's embrace rocks as our kin. We really do owe rocks everything because, without them, we wouldn't even be here. The elements that make up our bodies have been recycled through the lithosphere many times. The iron in our blood is the same iron found in the Earth's core. The food we eat relies on soil derived from rocks. The water in our bodies was once contained in magma, which erupted from a volcano and then circulated through the atmosphere, hydrosphere and biosphere. We cannot separate ourselves from our planet. Even the English word 'human' is derived from the reconstructed Proto-Indo-European root word **dhghem-*, meaning 'earth'. We are geological.

THE MAKING OF THE EARTH

Many geological processes happen over unimaginably long timescales that are hard to grasp. The Mid-Atlantic Ridge – which separates the Eurasian and North American tectonic plates – spreads at a rate of 2 to 5 centimetres (0.7 to 1.9 inches) per year. Sediments in the deepest parts of the oceans get thicker by 1 centimetre (0.4 inches) every 300 to 100,000 years. Granitic plutons – large underground chambers of hot magma – can take millions of years to cool down and crystallize. It takes a year for dead plants in bogs to form just 1 millimetre of peat. Because these processes happen so slowly, we often fail to notice them. In reality, they are shaping our world everywhere, all the time. Every second, a grain of sand moves in a river. Add billions of seconds together and the river repeatedly shifts course, floods, dries up and its sand turns to sandstone. Slowly but surely, the geological record is created.

Yet other processes happen in the blink of an eye. A volcanic eruption can last several days. A surging storm is over in a few hours. An earthquake shakes the ground in a matter of seconds. From a geological point of view, these processes are instant but equally able to make their mark.

Not all of these events get immortalized in rocks. Some evidence is eroded and forever lost to the depths of time. Every rock – or its absence – reveals a particular tale from the Earth's 4.55 billion-year history and transports us into a different version of our world. Through rocks, the slow ticking of geological time becomes a bit more tangible. By understanding rocks, we can develop a sense of timefulness.

The term timefulness was coined by geologist Marcia Bjornerud as a way to describe seeing the world as being made of time. The fact that our lifespan is tiny compared to the deep time recorded in rocks can either make us feel insignificant or we can embrace its immensity (I vouch for the latter). Everything that has ever happened led to now (it is why you're reading these words). In turn, now leads to everything that will ever happen in the future. Rocks allow us to witness this evolution. They are more than passive objects. You can think of rocks as verbs (doing), not nouns (being). They are actions, processes, events, changes and movements. Rocks actively participate in the making and shaping of the world.

ABOUT THIS BOOK

In this book, I decided to focus solely on rocks for two simple reasons. First, you are more likely to find a rock than a mineral. Second, it gives me the space to introduce you to a more diverse compendium of rocks from all over the world.

Each rock entry contains notes on its classification, origin or environment where it forms, size of its components, rocks it is easily confused with and the etymology of its name. Finally, there is a brief description that goes into more detail about how the rock forms and what geological processes it represents.

Most illustrations are based on real rocks I have encountered somewhere on my geological journey. Others are an amalgamation of a few different specimens of the same rock to help highlight its most important features. The diversity of rocks is incredible and because of that, it would be impossible to depict all the different varieties. I did my best to include a good range but can't claim that everything you will ever see is represented.

ROCK SPOTTING

Rocks are everywhere: from mountains, beaches and forests to kitchen countertops, walls and pavements. In contrast to many organisms that can only be seen at specific times of the year, rocks don't care about seasons or the weather. They are always there in one form or another. Thanks to their ubiquity, getting into rock spotting could not be easier – in fact it's harder to avoid rocks than it is to see them. But the key to understanding rocks is to look instead of see. The more rocks you look at, the easier it will be for you to identify them and recognize their geological narratives.

There is no right or wrong way to appreciate rocks. You can touch, paint, draw, sculpt, make pigments and glazes, research, write about, study the species that live on them, meditate upon, climb, walk across, dance on... The possibilities of learning with rocks, not just about them, are endless. Such an approach can help us see that they are more than inanimate resources to be exploited. It leads to a deeper way of knowing the world.

TOP TIPS

I like to research the geology of a place I am visiting before I go. That way, I know what to look for (and, I must admit, I often plan trips based on the rocks I would like to see). There are many local geological societies that have great resources and lots of geological maps are also available online, which makes the process much easier. When you are out and about, just keep your eyes peeled. No two rocks are the same, so you never know what you'll find.

You don't need to go far to see rocks. Exploring building materials in towns and cities can often reveal a lot about the bedrock they are built on. Historically, local rocks were used because they were accessible. Modern buildings often use rocks from farther away, so urban areas are perfect for seeing a wide variety of rocks in a relatively small space.

Counterintuitively, rock spotting isn't all about rocks. Rocks never occur in isolation. Their chemical composition influences the soils that form from them and the chemistry of the soil, in turn, supports a distinct community of organisms. Look at the plants, animals and

→
Hand lens
Perfect for observing the smaller details

fungi that live around and on rocks. Some plants (such as heather) thrive in the acidic soils on granites and gneisses, while others (like lavender and many meadow flowers) prefer the more alkaline soils atop limestones and chalks. Look at the shape of the land, too. For example, certain rocks are softer than others, so they weather and erode more easily. Noticing the wider landscape often reveals more about the geological bedrock than we might think.

You don't need any special equipment to appreciate rocks. However, the one bit of kit I would highly recommend (and I never leave the house without) is a hand lens. This pocket-sized magnifying glass is great for observing smaller details, like individual grains and crystals. Simply hold the hand lens close to your eye and move the rock you are observing as close as needed to get it in focus.

RESPECTING ROCKS

Rocks are our geological heritage, so it is important to preserve them for the future. It's not only about us, though. Many species directly depend on the protection rocks offer them, especially in rivers and on beaches, so any disturbance can negatively affect the ecosystem. Never hammer rocks from outcrops, stack pebbles into cairns or carve into rock faces. The best practice is always to leave no trace. However, if done responsibly and with care, collecting a rock or two is usually harmless. Make sure to familiarize yourself with local laws, as taking them may be forbidden, particularly on protected sites and Indigenous lands. If you can collect, don't hoard – you can always take photos.

LAYERS OF THE EARTH

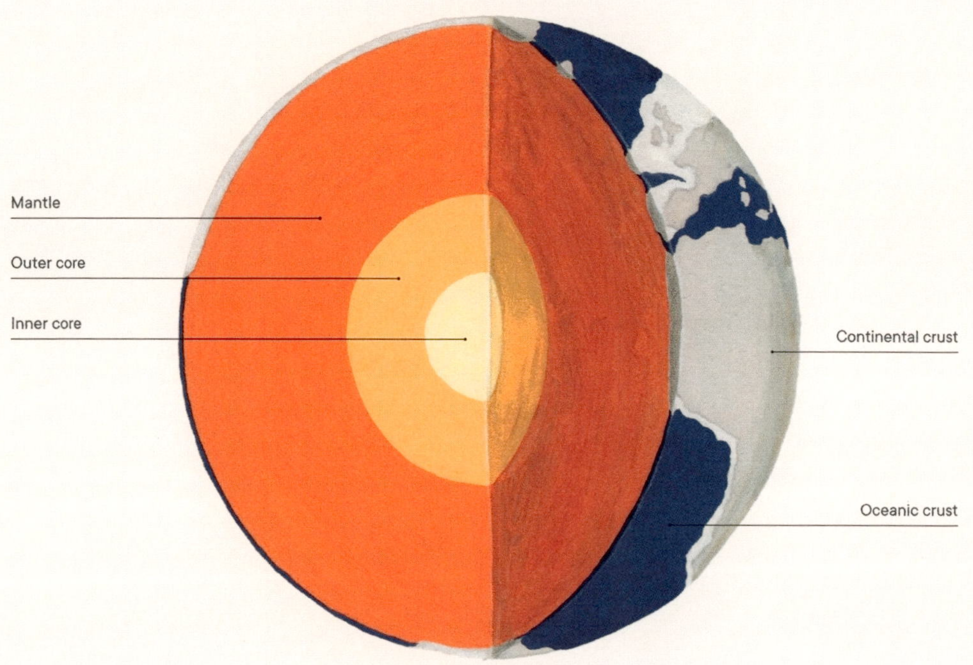

CORE

Nestled in the centre of the Earth is the core, made of a dense alloy of iron and nickel. It is split into two parts: the molten outer core and the solid inner core. The energy needed for geological processes is created here by the decay of radioactive elements. Movements inside the core also create the Earth's magnetic field.

MANTLE

Wrapping the Earth's core is a layer called the mantle. Contrary to popular belief, it is not liquid but mostly solid rock. And yet it moves. Over tens to hundreds of millions of years, great convective currents bring hot material from the core towards the crust and drive the restless movement of the tectonic plates above.

CRUST

The outermost layer – the one we live on – is the crust. There are two types: the thinner oceanic crust (6 to 11 kilometres or 3.7 to 6.8 miles thick) and the thicker continental crust (25 to 90 kilometres or 15.5 to 55.9 miles thick). It is split into tectonic plates that slowly move around.

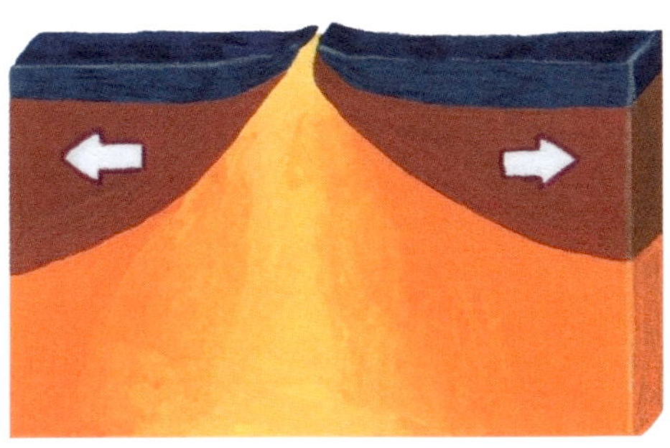

DIVERGENT BOUNDARY

When two plates move apart at divergent boundaries (rift valleys on continents and mid-ocean ridges in the oceans), new crust is created as hot magma from the mantle below rises to the surface and forms fresh rocks.

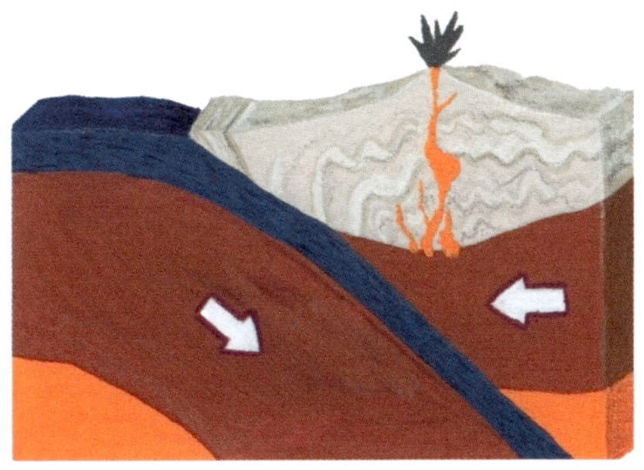

CONVERGENT BOUNDARY

Here, crust is destroyed in a process called subduction. When a cold, dense oceanic plate meets a less dense continental plate (or a younger, hotter oceanic plate), it slides under it. It carries water into the mantle, where it helps to melt the rocks of the overriding plate and creates a chain of mountainous volcanoes.

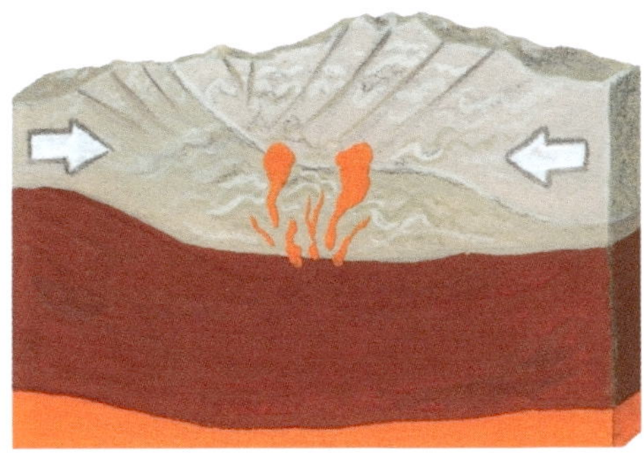

CONTINENTAL COLLISION

When two continental plates collide, neither wants to subduct and be melted in the mantle. They push against each other to raise mountain ranges during orogenic events. The intense pressures lead to large-scale folding and metamorphism of their rocks.

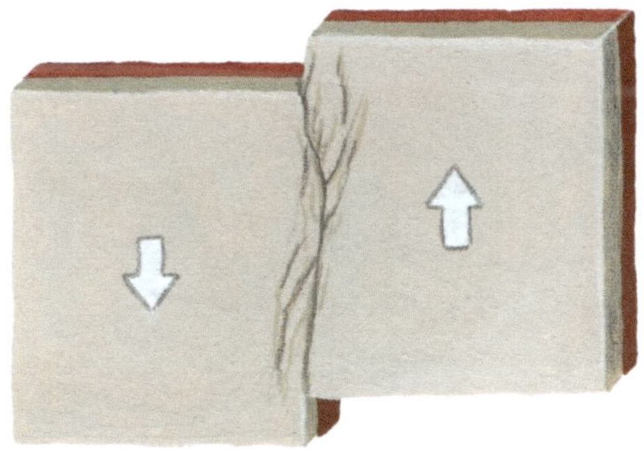

TRANSFORM BOUNDARY

Some tectonic plates slide horizontally past each other at transform (or conservative) boundaries. Here, crust is neither created nor destroyed. Instead, the friction caused by the grinding movement creates stress that often results in earthquakes.

MINERALS

Every rock is made of one or more minerals, and every mineral is a cocktail of various elements. There are thousands of different minerals but, fortunately for us, a few make a recurring appearance in most rocks. Most are silicates: their chemical structure is based on silica (SiO_2).

FELDSPAR

Feldspars are the most common mineral group in the Earth's crust. They share the same basic chemical formula, but each feldspar mineral contains a different proportion of calcium, sodium and potassium. Salmon-pink, yellowish, white or teal alkali feldspars are made of potassium and sodium. Milky-white to light-grey plagioclase feldspars have sodium and calcium. Their elongated crystals tend to be opaque or translucent.

QUARTZ

The composition of quartz is simple: oxygen and silicon. It is the second most abundant mineral in the crust and also one of the most resistant ones, which means it can be found in nearly all rocks. The best way to identify it is to look for translucent white to grey crystals. It is also commonly found filling mineral veins.

OLIVINE

Olivines are a group of minerals made of varying proportions of magnesium and iron. They are the first to crystallize in magma. Once they are exposed to the surface, olivine minerals weather very easily into different types of clays. Fresh olivines are yellowish green, while altered ones are brown to orange.

PYROXENE

Pyroxenes are another group of minerals made of magnesium and iron. They are divided into clinopyroxenes and orthopyroxenes based on their crystal structure (although they can really only be distinguished under a microscope). They tend to be a deep dark green to black and form shorter, stubby crystals. Pyroxenes are most often found in igneous and metamorphic rocks.

AMPHIBOLE

The name of this group is derived from the Latin *amphibolum* (ambiguous). There are so many different amphibole minerals that it is hard to distinguish between them. It also doesn't help that they look a lot like pyroxenes. Amphiboles are also dark green to black but tend to form hexagonal, needle-like crystals.

MICA

The two most common micas are dark biotite (rich in iron and magnesium) and light muscovite (rich in aluminium and potassium). They form shiny, sheet-like crystals that can be peeled apart (they are sometimes called mica books). Micas are quite resistant to erosion, meaning they can be found in almost all rock types.

GARNET

Yellow to red (and sometimes green) garnets are easy to spot: their dodecahedral crystals make them look like the seeds of pomegranates, which is how they got their name. Garnets can only form under high pressure and high temperature deep in the Earth, so are mostly found in metamorphic rocks.

FELDSPATHOID

Feldspathoids (also shortened to foids) want to be feldspars but don't have enough silica for it. Instead, they use more aluminium and potassium in their structure. Because of the lack of silica, feldspathoids never form in the same rock as quartz, but they can be found alongside feldspars.

CALCITE

Calcite is the only main mineral that is not based on silica, but instead on carbonate (CO_3). It is most commonly found in sedimentary rocks but can also be present in igneous and metamorphic rocks. Like quartz, it often forms in mineral veins.

THE ROCK CYCLE

Rocks are constantly being changed by the processes happening inside the Earth and on the surface. They are far from stable. Every rock is always becoming a different rock.

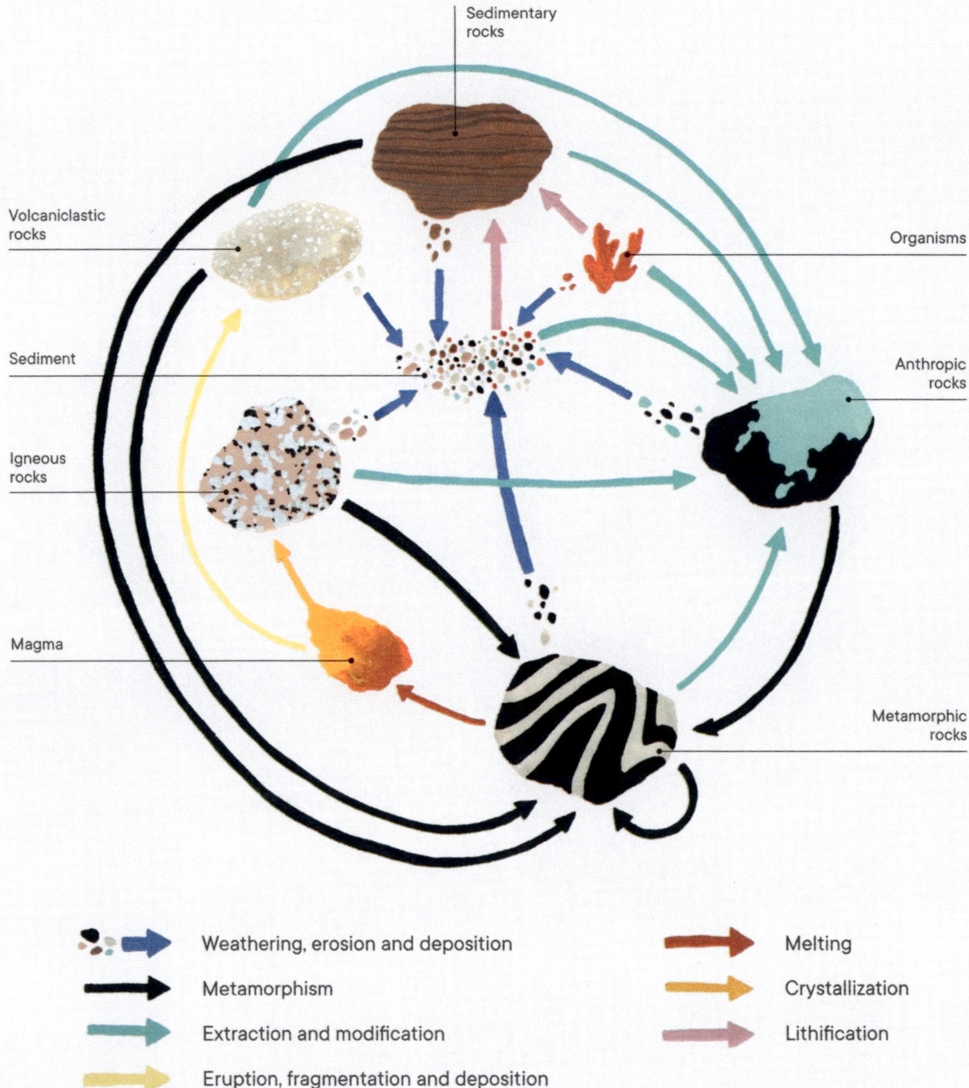

IDENTIFYING THE MOST COMMON ROCKS

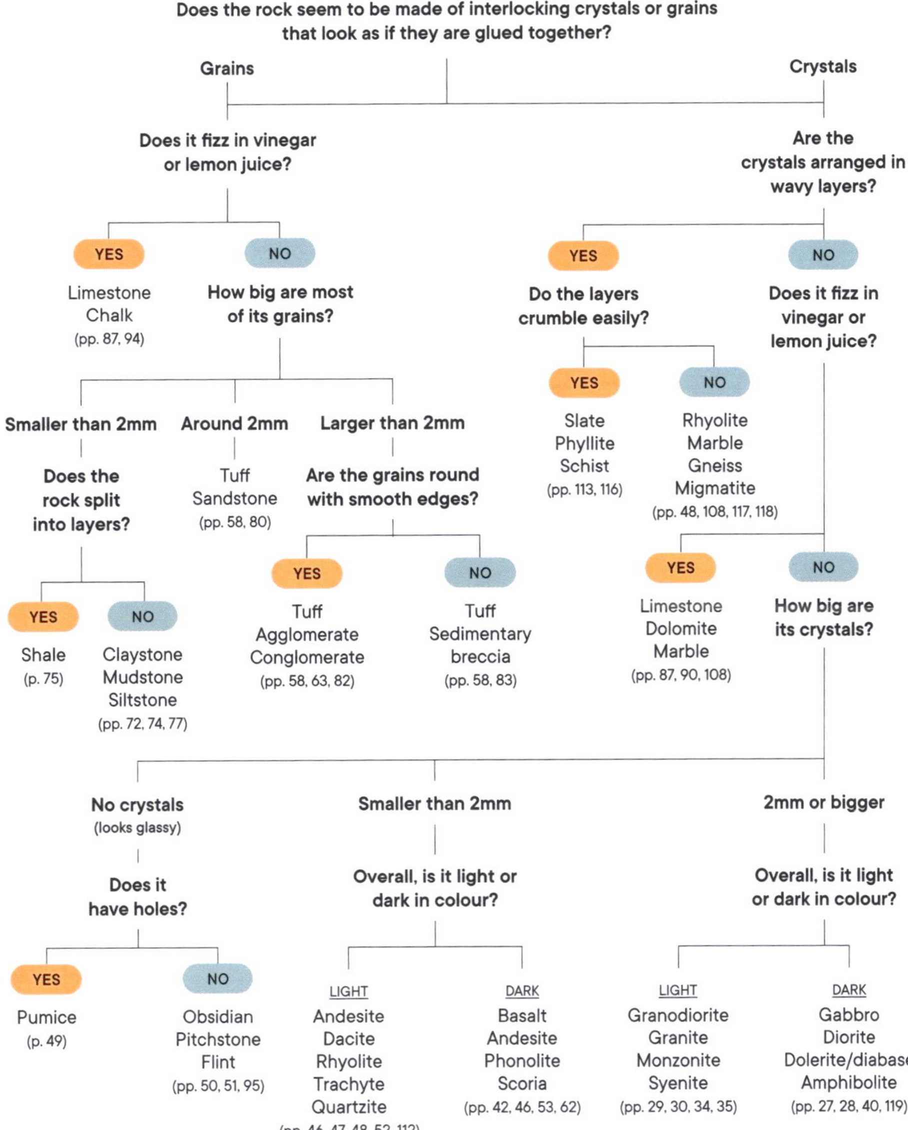

IGNEOUS

IGNEOUS ROCKS

From Latin *ignis* (fire)

When our planet formed around 4.55 billion years ago, her surface wasn't really a surface, but a vast ocean of magma. Within a hundred million years or so, the Earth cooled just enough to create islands of solid rock and actual oceans filled with water (great news for any early life forms). These early continents were formed from igneous rocks, which were created when magma cooled down and crystallized. Since then, igneous rocks have experienced a boom (some of them literally) and form in all sorts of places.

Plutonic (also called intrusive or abyssal) rocks slowly crystallize underground in magma reservoirs. This gives their minerals enough time to grow large (above 2 millimetres).

Subvolcanic (also called intrusive or hypabyssal) rocks crystallize below the surface in shallow intrusions. The mineral grains are normally medium-sized (0.25 to 2 millimetres).

Volcanic (also called extrusive) rocks erupt from volcanoes as lava (during effusive eruptions) or ash (during explosive eruptions). As they come into contact with air, water or ice, they cool down rapidly, which doesn't give any minerals enough time to grow large. Volcanic rocks have a groundmass made of fine mineral grains (less than 0.25 millimetres) or no grains at all in the case of volcanic glass.

Depending on their chemical composition, we can distinguish between ultramafic, mafic, intermediate and felsic igneous rocks. You can imagine them as a spectrum: ultramafic and mafic rocks have virtually no quartz (silica), while felsic rocks are full of it. Intermediate rocks sit somewhere between them. If the specific rock also has an abundance of sodium and potassium, it is classified as alkaline.

Before an igneous rock forms, it needs to be molten. Magma that will crystallize into an igneous rock can form via **partial melting** or **fractional crystallization**, or a combination of the two.

At mid-ocean ridges, the nearly solid mantle is so close to the surface that it decompresses and starts to partially melt. This creates basaltic magma, which forms new oceanic crust. Partial melting also occurs in subduction zones, where an oceanic plate carries water down into the mantle. While the mantle was more or less solid until that point, the water reduces its melting temperature. The molten mantle material rises until it meets rocks of the overriding plate. Their felsic minerals (quartz and alkali feldspar; represented in grey in the illustrations below) melt first, while the more temperature-resistant mafic minerals (olivine and pyroxene; black) stay behind. The felsic magma (orange) is less dense than the surrounding rock, so it rises and builds more continental crust.

During fractional crystallization, we start with a molten magmatic broth. Mafic minerals (represented in black in the illustrations below) crystallize first, removing magnesium and iron from the melt, and incorporating them into their structure. The magma (orange) evolves. Felsic minerals (grey) start growing using the leftover calcium, sodium, potassium and silica. Depending on how much magma is available and its initial composition, this process creates a variety of igneous rocks with different chemical compositions.

Some igneous rocks include bits of other rocks that are mineralogically not the same, called xenoliths (from Greek *xénos* – a guest – and *líthos* – stone). These 'guest rocks' get torn from the surrounding bedrock as magma rises through the crust. Sometimes, the rock gets absorbed into the magma apart from a few of its crystals – such 'guest crystals' are then called xenocrysts.

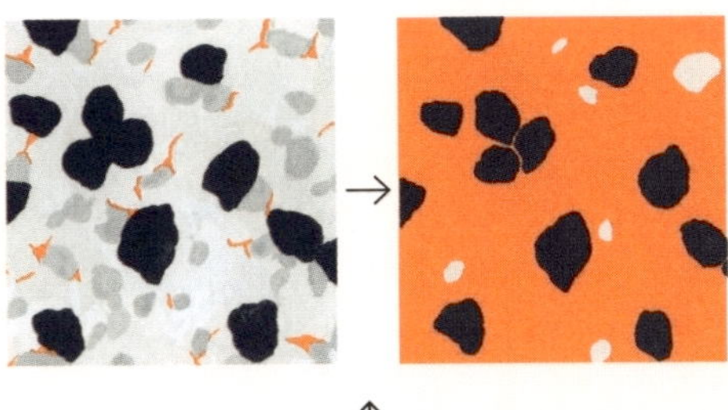

↑

Partial melting

Existing solid rock melts to create new magma

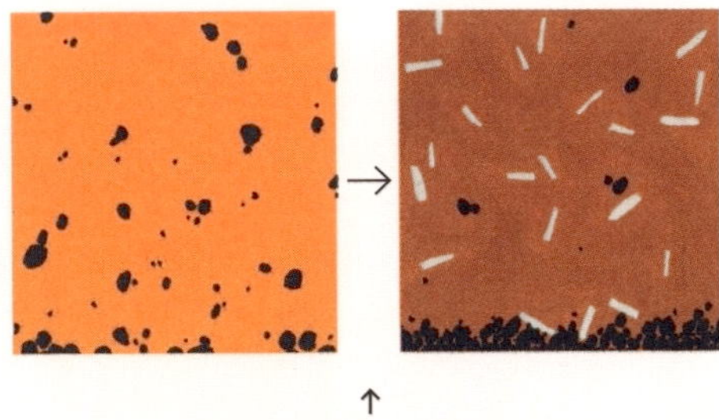

↑

Fractional crystallization

Different minerals crystallize from magma in stages

AMYGDALOIDAL ROCK

BUBBLES FILLED WITH MINERALS

Classification	Mafic to felsic (can be alkaline)
Origin	Volcanic
Grain Size	Fine
Similar To	Any volcanic igneous rock
Name Origin	From Ancient Greek *amygdalē* (almond)

All magmas contain gases and liquids (together called volatiles), most commonly water and carbon dioxide. In the depths of magma chambers, they are happily dissolved because of the pressure of the rock above. But when magma erupts and the pressure decreases, these volatiles separate from the molten rock, form into bubbles and try to escape. Not all of them can break free, though, and some get trapped. As the lava around them cools down, they literally freeze in place, leaving behind round to elongated holes called vesicles. If mineral-rich hydrothermal fluids flow through these gaps, they can be filled by secondary minerals like quartz (including agate and jasper), calcite, chlorite or zeolites. Such infilled vesicles are called amygdales and the rock (most often basalt or andesite) amygdaloidal. This sample of andesite is both amygdaloidal and porphyritic.

PORPHYRY

MAGMA COOLED IN STAGES

Classification	Mafic to felsic (can be alkaline)
Origin	Plutonic, subvolcanic, volcanic
Grain Size	Fine to coarse with larger crystals
Similar To	Any igneous rock
Name Origin	From Ancient Greek *porphyrítēs líthos* (purple stone)

Porphyry is a term for any igneous rock with larger crystals (called phenocrysts) floating in its groundmass. Such texture is caused when magma cools in stages: the phenocrysts slowly crystallize inside the magma chamber until something forces the rest of the molten rock to cool down more quickly. This most often happens during an eruption or a shallow intrusion. The sudden cooling forces the rest of the melt to crystallize almost instantly, giving the minerals less time to form larger phenocrysts. Pretty much any igneous rock can be porphyritic. The porphyry that started it all, though, was a dacite quarried by the Romans in Mons Porphyrites in Egypt. This rock was highly prized for its distinct purple colour; it was called Imperial Porphyry and was used as a status symbol by Roman Emperors. See also amygdaloidal rock, lamprophyre (p. 37), andesite (p. 46), dacite (p. 47), pitchstone (p. 51), trachyte (p. 52), phonolite (p. 53) and the cobble on p. 68 for more examples of porphyries.

PEGMATITE

WHO HAS THE LARGEST CRYSTALS OF THEM ALL?

Classification	Mafic to felsic (can be alkaline)
Origin	Plutonic
Grain Size	Very coarse
Similar To	Granite, syenite, gabbro
Name Origin	From Ancient Greek *pêgma* (something joined together)

Pegmatite is a term for a plutonic rock of any composition with exceptionally large crystals (normally around 1 to 5 centimetres or 0.4 to 2 inches, but colossal 18-metre or 15-foot crystals are also known). They are most often found near the edges of larger intrusions, sometimes extending into the surrounding bedrock as veins. Pegmatites form from the very last dregs of magma after the rest of the intrusion has cooled down and crystallized. These remaining pockets are full of water, which allows the minerals to grow to their incredible sizes at the rapid rates of a few centimetres per year. This melt is also rich in the so-called incompatible elements that don't like being part of any of the main minerals. They hold off until the very last moments to crystallize. A lot of pegmatites thus contain rare minerals and gemstones made of these elements, like corundum (and its varieties ruby and sapphire), beryl, topaz and tourmaline.

1 Granite pegmatite

2 Syenite pegmatite with corundum

1 Lherzolite xenoliths in basalt
2 Dunite with chromite

PERIDOTITE

VISITORS FROM THE DEPTHS

Classification	Ultramafic
Origin	Plutonic
Grain Size	Medium to coarse
Similar To	Troctolite, gabbro
Name Origin	After peridot (a gem variety of the mineral olivine) which is borrowed from French *péridot*

Peridotite refers to a bunch of greenish rocks that form deep within the mantle. Because they make up the inside of the Earth, they are rare on the surface. Small bits of peridotite are sometimes trapped in basaltic magma that rises to the surface and then get erupted as xenoliths – guest rocks – in lava. The Earth's surface is a very different place compared to the hot, pressurized mantle. It is a downright extreme environment for peridotites, so they tend to quickly weather into brownish-orange rocks. They can be divided based on the proportions of their two minerals into pyroxenite (both pyroxenes), dunite (pure olivine), wehrlite (olivine + clinopyroxene), harzburgite (olivine + orthopyroxene) and lherzolite (olivine + both pyroxenes).

ANORTHOSITE & TROCTOLITE

MOON ROCKS

Classification	**Mafic**
Origin	**Plutonic**
Grain Size	**Medium to coarse**
Similar To	**Gabbro**
Name Origin	**Anorthosite after its main mineral, the feldspar anorthite, whose name is derived from Greek *án* (not) + *órthós* (straight) because of its uneven crystal angles; troctolite from Greek *trōktēs* (a sea fish, trout) for its spotty appearance**

1

Next time you look up at the Moon, notice the pale white lunar highlands – they are made of anorthosite and troctolite. Back on Earth, these rocks form in large intrusions during fractional crystallization from magma. The first mineral to crystallize is olivine and it gets removed from the melt as it settles at the bottom of the magma chamber due to its higher density. This slightly changes the composition of the remaining melt and leads to the crystallization of pyroxene and plagioclase feldspar. Their settling creates distinct layers of different rocks: at the bottom is the olive-green peridotite (olivine + pyroxene), followed by green-white troctolite (olivine + pyroxene + plagioclase feldspar) and capped by white anorthosite (plagioclase feldspar).

2

1 Anorthosite

2 Troctolite

GABBRO

BELOW THE OCEAN FLOOR

Classification	**Mafic**
Origin	**Plutonic**
Grain Size	**Medium to coarse**
Similar To	**Dolerite, diorite**
Name Origin	**After its type locality near the village Gabbro in Italy or perhaps from classical Latin *glaber* (smooth)**

Gabbro is found everywhere beneath the Earth's oceanic crust. As hot magma upwells from the mantle at mid-ocean ridges, it creates large gabbroic reservoirs that in turn feed the eruptions of basalt at the ocean floor. On land, it can be found in exposed ophiolites – portions of the oceanic crust that got uplifted onto land by tectonic forces. It is also known in layered igneous intrusions where it forms during fractional crystallization after troctolite and anorthosite, or as eroded magma chambers that supplied basaltic magma for ancient eruptions. Gabbro is made of large, dark pyroxene crystals surrounded by white-to-grey calcium-rich plagioclase feldspar.

DIORITE

THE MOTHER MAGMA OF ANDESITE

Classification	**Intermediate**
Origin	**Plutonic**
Grain Size	**Medium to coarse**
Similar To	**Gabbro, granodiorite**
Name Origin	**From Greek *diorízein* (to distinguish)**

Visually, this rock is easy to confuse with gabbro. The main difference is in the chemical composition of their plagioclase feldspar minerals: those in diorite are rich in sodium as opposed to the calcium in gabbro. Diorite also contains dark hornblende (an amphibole mineral), giving it a salt-and-pepper-like appearance. It is a rock typically found in magma chambers above subduction zones where it feeds eruptions of andesitic lavas. It often forms batholiths – large, complex intrusive bodies that can cover areas of over 100 square kilometres (38.6 square miles). Diorite can also occur alongside granite and gabbro as it lies compositionally between them.

TONALITE, TRONDHJEMITE & GRANODIORITE

ROOTS OF THE OLDEST CONTINENTS

Classification	Intermediate to felsic
Origin	Plutonic
Grain Size	Medium to coarse
Similar To	Diorite, granite
Name Origin	Tonalite after its type locality in Passo del Tonale in Italy; trondhjemite after its type locality near Trondheim in Norway; granodiorite because it is compositionally between granite and diorite

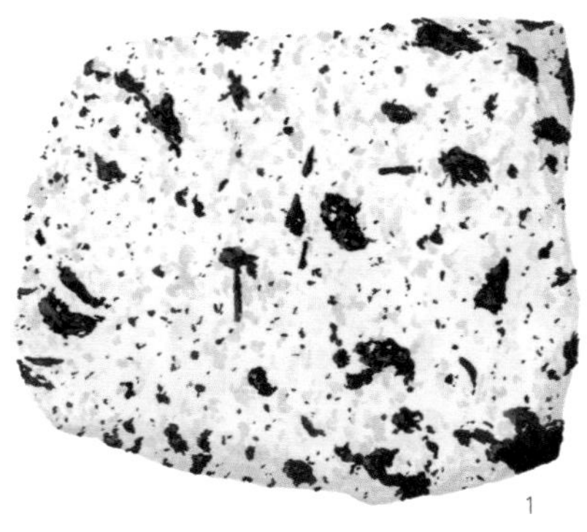

1

Tonalite, trondhjemite and granodiorite are so often found together that they are usually shortened to 'the TTG suite' to make everyone's life easier. These rocks formed the Earth's first continental crust during the Archaean Eon (4 to 2.5 billion years ago). They were likely made by partially melting the ancient ocean floor (though these oldest examples have since metamorphosed into gneisses). All three have a very similar composition of plagioclase feldspar, quartz, some dark minerals (usually biotite, sometimes hornblende) and a little bit of alkali feldspar. Tonalite contains only up to 10% alkali feldspar; trondhjemite – a variety of tonalite – is distinguished by the presence of plagioclase feldspar rich in sodium. Granodiorite does have a bit more alkali feldspar but still not enough to make it a granite.

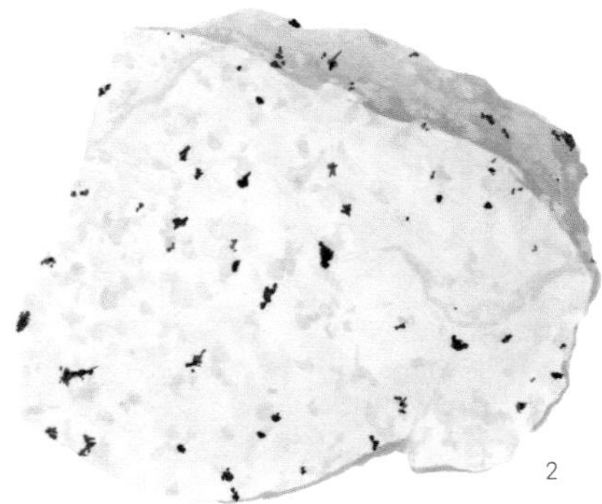

2

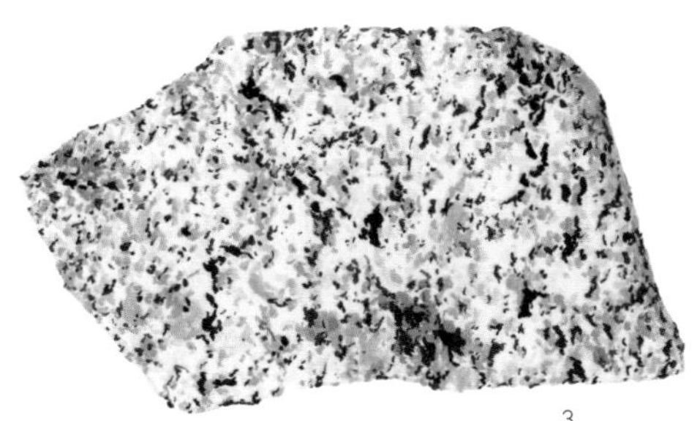

3

1 Tonalite

2 Trondhjemite

3 Granodiorite

GRANITE

THE MOST COMMON PLUTONIC ROCK

Classification	Felsic
Origin	Plutonic
Grain Size	Medium to coarse
Similar To	Granodiorite, syenite
Name Origin	From Latin *granum* (grain)

Granite is the most abundant plutonic rock in the Earth's continental crust. It usually forms above active subduction zones where oceanic plates carry water down into the mantle.
The addition of water allows the pre-existing metamorphic rocks of the overriding plate to partially melt: their felsic minerals melt first while the mafic minerals stay solid (see migmatite, p. 118). The resulting magma is less dense than the surrounding bedrock, so it rises and collects in large underground stocks. There, it slowly cools

down and crystallizes into granite over millions of years. However, this process is extremely complex and can differ from one stock to another – this is why granites come in so many colours. They share a mineral makeup of quartz, alkali feldspar and plagioclase feldspar, plus varying amounts of biotite, muscovite and hornblende. Some granites also contain garnets and/or tourmaline.

Graphic granite is a type of pegmatite (p. 24) that contains only two minerals: quartz growing within alkali feldspar. Its name was inspired by its resemblance to ancient cuneiform writing. If these intergrowths are microscopic, the rock is called granophyre.

1 Muscovite granite
2 Graphic granite
3 Alkali feldspar granite
4 Granite with alkali feldspar, plagioclase feldspar, quartz and biotite

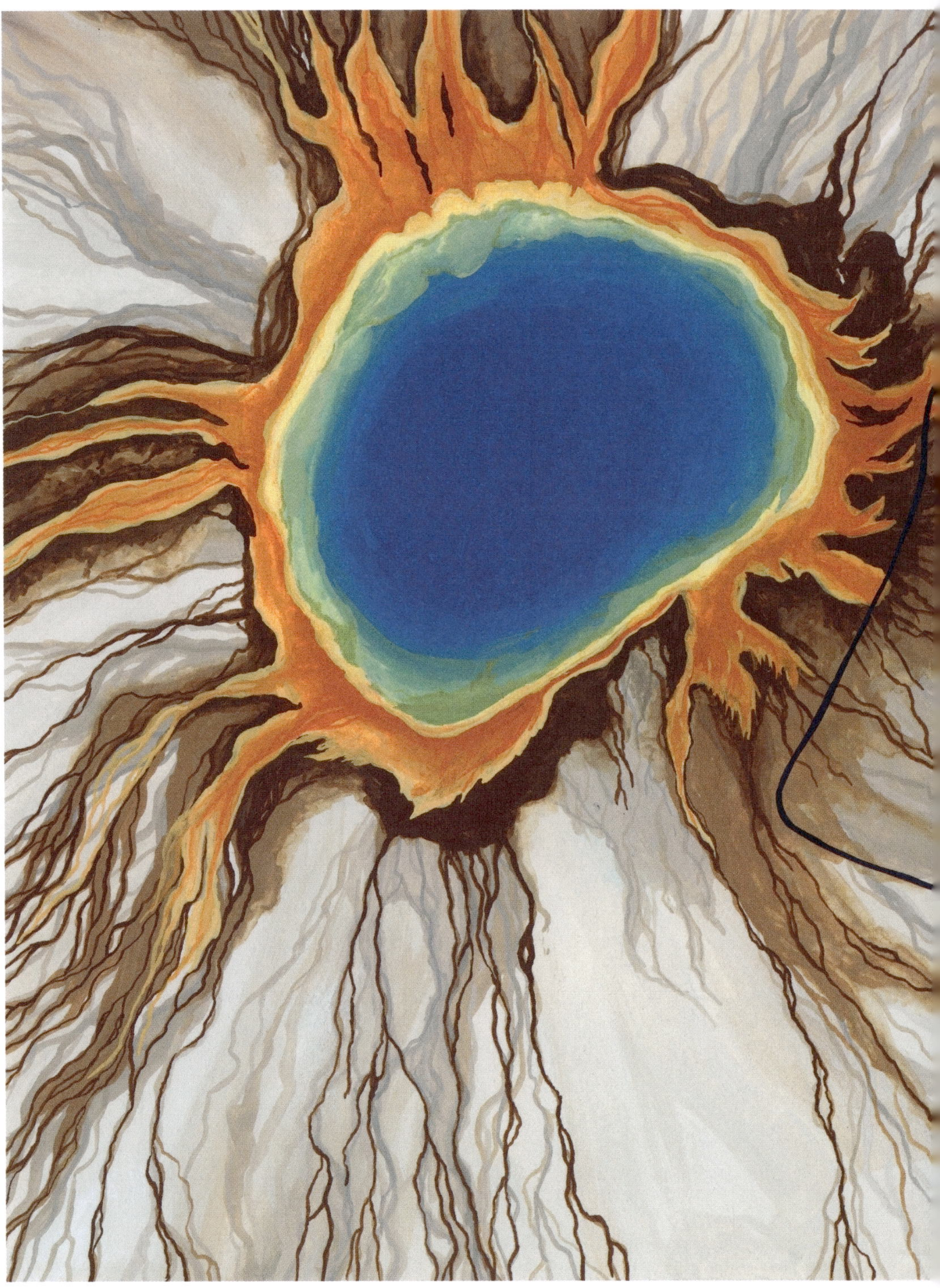

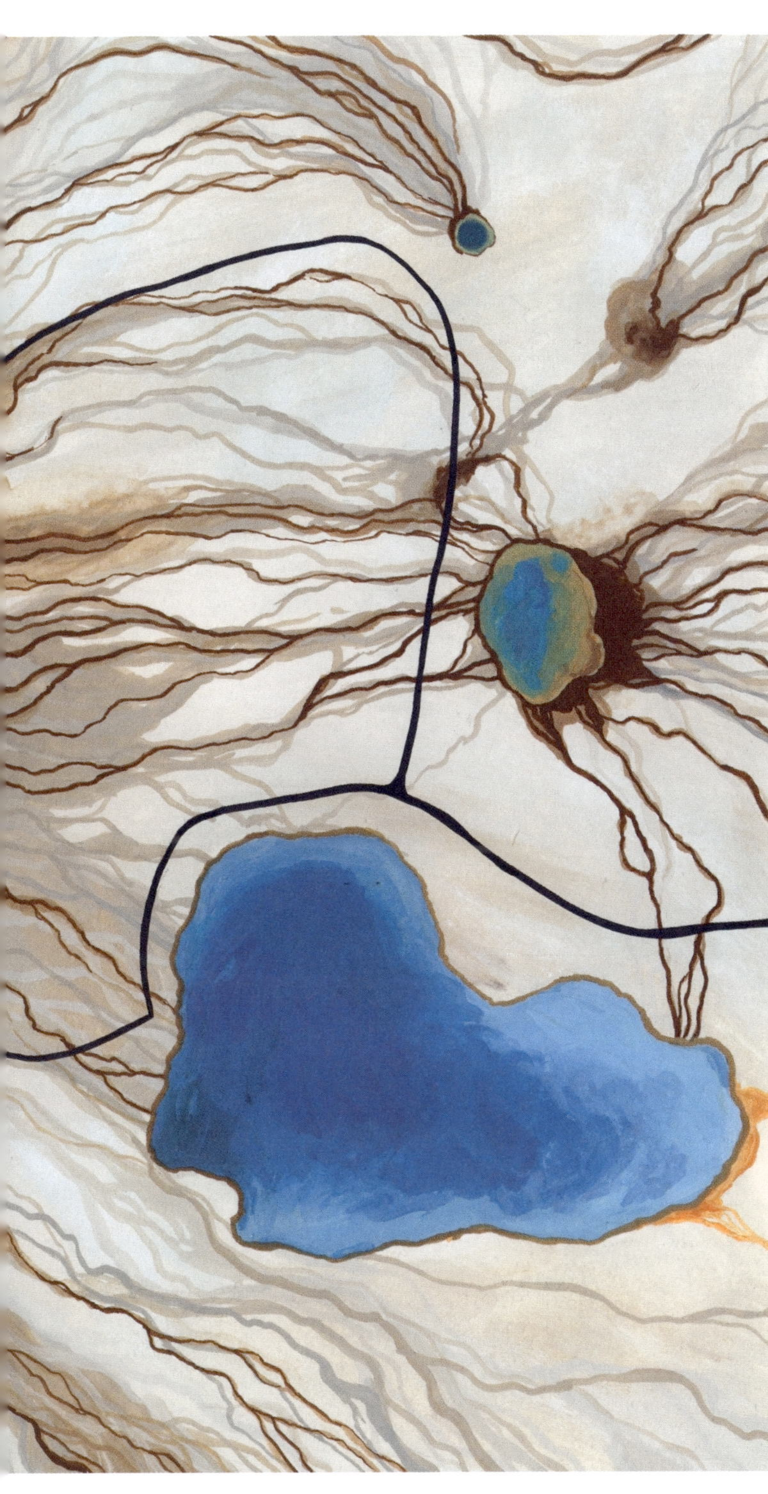

←
Yellowstone Caldera
Wyoming, USA

Today, granite is slowly crystallizing in the cooling magma chamber beneath the Yellowstone Caldera. It last erupted lava around 70,000 years ago. The mostly solid granite still gives off enough energy to heat groundwater for powerful eruptions of geysers and hot volcanic springs on the surface.

MONZONITE & LARVIKITE

ALL THE FELDSPARS

Classification	**Intermediate (alkaline)**
Origin	**Plutonic**
Grain Size	**Medium to coarse**
Similar To	**Diorite, syenite**
Name Origin	**Monzonite after its type locality in the Monzoni Range in Italy; larvikite after its type locality near Larvik in Norway**

A fairly uncommon rock, monzonite probably crystallizes from the same magma as granite and is often found in smaller intrusions surrounding it. It looks like granite too – the only difference is its distinct absence of quartz. Monzonite contains equal parts of alkali feldspar and plagioclase feldspar, which give it a pink and white mottled appearance. Some dark minerals (like hornblende and biotite) are also often present.

Larvikite, a variety of monzonite, is known for its bluish iridescent sheen called the schiller effect. It is caused by the unique chemical composition of its feldspar crystals: unusually, they contain all calcium, sodium and potassium in their structure.

1 Monzonite
2 Larvikite

SYENITE & FOID SYENITE

ALMOST GRANITE

Classification	Intermediate (alkaline)
Origin	Plutonic
Grain Size	Medium to coarse
Similar To	Monzonite, granite
Name Origin	After the Egyptian city Syene (Aswan) where it was quarried in ancient times

The best way to distinguish this rock from granite is to look for the bluish translucent crystals of quartz – there aren't any in syenite. It also has more alkali feldspar than plagioclase feldspar (a good way to tell it apart from monzonite), plus the dark minerals hornblende, pyroxene and biotite. Much like granite, syenitic magma is thought to form through partial melting. When metamorphic rocks are exposed to high enough pressure and temperature deep in the Earth, parts of them melt to form magma (see migmatite, p. 118).

Foid syenites contain feldspathoid minerals on top of the classic minerals found in syenite – most often the cloudy white nepheline or bluish sodalite. The Greenlandic kakortokite (named after its type locality near Qaqortoq) has well-defined contrasting layers, each dominated by either black arfvedsonite, pink eudialyte or white nepheline. This distinct layering formed when crystals settled in stages at the bottom of an ancient magma chamber.

1

2

3

1 Biotite syenite 2 Kakortokite 3 Hornblende syenite

ORBICULITE

THE BEST-LOOKING PLUTONIC ROCK

Classification	**Mafic to felsic (can be alkaline)**
Origin	**Plutonic**
Grain Size	**Medium to coarse**
Similar To	**Gabbro, diorite, granodiorite, granite**
Name Origin	**From classical Latin *orbiculus* (a small ball)**

Orbiculite is definitely one of the more unusual igneous rocks. It is immediately recognizable by its spheroid shapes called orbicules. These magmatic balls are made of concentric layers of elongated minerals (usually plagioclase feldspar and hornblende) but it's not entirely clear how they form. Something must change in the cooling magma chamber that forces the minerals to crystallize into spheres. They can then float around the magma chamber and concentrate in a place where they won't melt again, which is why orbiculite is usually found covering only a relatively small area. The rock surrounding the orbicules can have various compositions, including gabbro, diorite, granodiorite and granite.

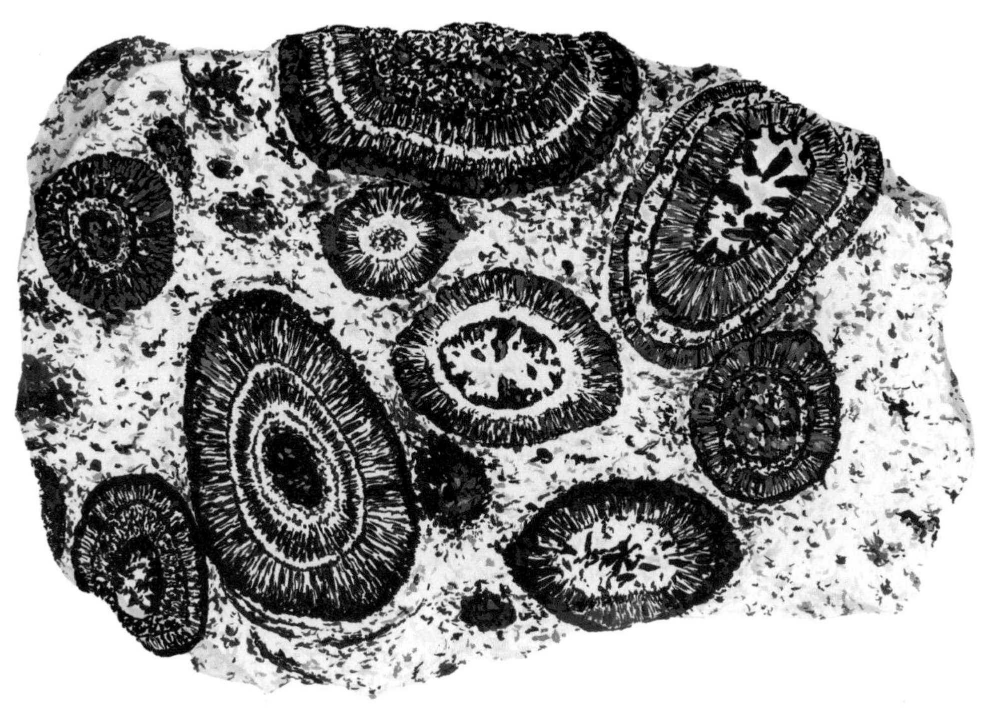

LAMPROPHYRE

THE MOST CONFUSING IGNEOUS ROCK

Classification	Who knows?
Origin	Subvolcanic
Grain Size	Fine
Similar To	Dolerite, kimberlite
Name Origin	From Greek *lámpein* (to shine) + *phanerós* (visible)

Lamprophyres are a bunch of mysterious rocks that were grouped together by early geologists because they all looked more or less the same and were often found as intrusions. There are a lot of varieties named after the places where they were first discovered, such as minette (France), camptonite (USA) and monchiquite (Portugal, pictured). However, these varieties may not even have anything in common, chemically speaking. They usually (but not always and not all) contain biotite, amphibole, pyroxene, plagioclase feldspar, alkali feldspar and/or feldspathoids. Some minerals can be present as large glistening phenocrysts, which gave lamprophyres their name.

CARBONATITE

THE STRANGEST IGNEOUS ROCK

Classification	**Carbonate**
Origin	**Plutonic, subvolcanic, volcanic**
Grain Size	**Fine to coarse**
Similar To	**Limestone**
Name Origin	**From Latin *carbō* (charcoal) as it contains carbonate minerals**

Carbonatite is probably the most bizarre igneous rock because its main minerals are not silicates but carbonates (the same minerals found in limestone, p. 87). The source of carbonatite magma is a mystery. It is often associated with alkaline igneous rocks and is thought to come from the mantle, but no one is entirely sure how exactly it forms. Older carbonatites are found all around the world, but the only currently active carbonatite volcano is Ol Doinyo Lengai in Tanzania. It produces cold (below 600°C or 1,110°F), very fluid lava with only a faint glow. The lava is dark greywhen liquid but, because of the water-soluble carbonate minerals it contains, it quickly alters into a whitish rock once cool.

KIMBERLITE

THE FASTEST-MOVING ROCK

Classification	Ultramafic
Origin	Subvolcanic or volcanic
Grain Size	Variable
Similar To	Peridotite, lamprophyre
Name Origin	After its type locality near Kimberley in South Africa

Forming some 400 kilometres (250 miles) below the surface, kimberlite is one of the deepest mantle rocks. We know it forms so deep because it sometimes contains the hardest material on Earth: diamonds. They can only form under extremely high pressure in the mantle and, as soon as the pressure drops, diamonds turn into their more stable form, graphite. To reach the surface, they need speed. Kimberlite rises to the surface rapidly, tearing away bits of any rocks it encounters (most often peridotite, p. 25). The rest of the rock is made of serpentinized olivine, pyroxene, phlogopite (a type of mica) and sometimes garnet. On the surface, it forms characteristic shallow, carrot-shaped intrusions and volcanic craters called diatreme pipes.

DOLERITE/DIABASE

A ROCK WITH MANY NAMES

Classification	Mafic
Origin	Subvolcanic
Grain Size	Medium
Similar To	Gabbro, basalt
Name Origin	Dolerite from Greek *dolerós* (deceptive); diabase from Ancient Greek *diábasis* (the act of crossing over)

This rock changes name based on your location: it's diabase in North America and dolerite in the rest of the English-speaking world. We could also call it microgabbro because, exactly like gabbro, it is made of plagioclase feldspar and pyroxene, the crystals are just smaller. Dolerite usually forms small bodies cutting through other rocks: shallow intrusions around ancient volcanoes like vertical dykes (or dikes, if you prefer the North American spelling) and horizontal sills (that's the only spelling, thank goodness) are the most common. It is also found below the ocean floor where it acts as pipes that carry magma towards the mid-ocean ridge.

KOMATIITE

AN EXTINCT ROCK

Classification	Ultramafic
Origin	Volcanic
Grain Size	Fine
Similar To	Basalt
Name Origin	After its type locality near the Komati River in South Africa

At up to 3.8 billion years old, this rock formed the oldest lavas on Earth before it stopped erupting around 2.5 billion years ago. The reason komatiite lavas aren't flowing anymore is because the Earth's mantle is slowly cooling down. Komatiitic magma needs temperatures of at least 1,400°C (2,550°F) to form but today the mantle is only around 1,200°C (2,190°F). Because of the high temperature, komatiite lava flows would have had very low viscosity (similar to water) and flowed rapidly as thin sheets over long distances. As it cooled, the olivine grew into blade-like crystals with a feathery texture called spinifex (named after a species of grass).

BASALT

THE MOST COMMON VOLCANIC ROCK

Classification	Mafic
Origin	Volcanic
Grain Size	Fine
Similar To	Dolerite, andesite
Name Origin	A misspelling of Greek *básanos* (touchstone, a trial), which came from the Egyptian *baban* (slate used as a touchstone to test gold)

Basalt is the most abundant volcanic rock on Earth. It makes up the bulk of oceanic crust and has been erupting throughout geological time all over the planet. It is made of plagioclase feldspar rich in calcium, pyroxene and sometimes olivine – it is the volcanic equivalent of gabbro and dolerite (diabase). Basaltic magma is hot (around 1,200°C or 2,190°F) and has a low viscosity (about the same as honey), making it easy for any volatile gases to escape. Because of that, most basaltic eruptions are effusive, producing either ropy (called by its Hawaiian name pāhoehoe) or blocky (called 'a'ā) flows that create flat and wide shield volcanoes.

RETICULITE, TACHYLYTE, PELE'S TEARS, PELE'S HAIR & LIMU O PELE

BASALTIC GLASS

Classification	Mafic
Origin	Volcanic
Grain Size	Fine
Similar To	Scoria, pumice
Name Origin	Reticulite from classical Latin *rēticulum* (a small net); tachylyte from Greek *takhús* (swift) + *lutós* (loose, dissolved)

Glass forms when basaltic lava cools down very quickly. The most vigorous basaltic eruptions create fountains that throw lava high in the air. As the lava breaks into pieces, any gases it contains are released into the atmosphere and leave behind tiny holes called vesicles. Reticulite (or thread-lace scoria) is a kind of basaltic pumice made of a delicate network of these gas bubbles. Tachylyte (also spelled tachylite) forms during underwater eruptions; it is often found as rims on pillow lavas. Some distinct shapes of glass are named after Pele, the Hawaiian goddess of volcanoes. Pele's tears are droplets that got their aerodynamic shape as they fell to the ground, while Pele's hair is formed of thin glassy strands often carried by wind. Limu o Pele (Pele's seaweed) is created when lava flows are rapidly cooled down and burst into giant bubbles as they enter the ocean.

1 Reticulite
2 Tachylyte
3 Pele's tears
4 Pele's hair
5 Limu o Pele

←
Reykjanes Peninsula
Iceland

The volcanic system beneath Reykjanes became active in late 2019 after almost 800 years of slumber. Magma made its way to the surface in a series of intrusions, each accompanied by a swarm of earthquakes. It finally erupted in March 2021 and covered the mossy earth of Iceland in fresh basaltic lava.

ANDESITE

THE MOST COMMON VOLCANIC ROCK ON CONTINENTS

Classification	**Intermediate**
Origin	**Volcanic**
Grain Size	**Fine**
Similar To	**Basalt, dacite**
Name Origin	**After the Andes Mountains where it was first described; the Andes themselves take their name from the Quechua *anti* (high crest)**

Andesite consists of sodium-rich plagioclase feldspar, hornblende and pyroxene; all can be present as larger phenocrysts formed while the magma was chilling underground. Depending on their ratios, the rock can be light to dark grey. These minerals contain more silica, which makes andesitic lava more viscous – it flows about as easily as smooth peanut butter (or crunchy if we consider the phenocrysts). Because andesitic magma is produced above subduction zones, it is also full of water and gases that can't always easily escape during an eruption. As a result, it erupts both effusively and explosively. Stratovolcanoes, made of layers of lava and ash, are the ideal places to see this rock.

DACITE

A TOWER-BUILDING ROCK

Classification	Intermediate to felsic
Origin	Volcanic
Grain Size	Fine
Similar To	Andesite, rhyolite
Name Origin	After its type locality in Dacia, an ancient kingdom in present-day Romania and Moldova

Dacite shares its mineral make-up of plagioclase feldspar, quartz and some alkali feldspar with granodiorite. Biotite, hornblende and pyroxene are also sometimes present, both in the groundmass and as phenocrysts (the sample illustrated has distinct quartz, plagioclase feldspar and biotite). The higher amount of silica present in quartz makes dacitic magma very viscous, which produces explosive eruptions of ash. It can erupt effusively, too, but the lava tends to build towering domes and spines that block the volcanic crater. This dacitic plug puts an immense amount of pressure on the already pressurized magma below, which can trigger an even more explosive eruption.

1 Rhyolite
2 Flow-banded rhyolite

RHYOLITE

THE MOST EXPLOSIVE LAVA

Classification	Felsic
Origin	Volcanic
Grain Size	Fine
Similar To	Andesite, dacite
Name Origin	From Ancient Greek *rhýax* (stream, flow)

Rhyolite – the volcanic equivalent of granite – is found erupting from volcanoes above subduction zones. It is the most evolved volcanic rock. The groundmass is usually lightly coloured and glassy, sometimes with phenocrysts of alkali feldspar and quartz. Because of its high amount of silica, rhyolitic lava has a low temperature (around 900°C or 1,650°F) and is extremely viscous. The individual silica molecules in quartz form chains that prevent it from moving easily. Add a high amount of volatiles (water and gases sourced from the subducting oceanic crust) and we get highly explosive eruptions of ash. If the eruption is effusive, rhyolitic lava doesn't flow very far and is often banded.

PUMICE

THE ONLY ROCK THAT FLOATS ON WATER

Classification	Intermediate to felsic (can be alkaline)
Origin	Volcanic
Grain Size	Glassy
Similar To	Scoria
Name Origin	From classical Latin *pūmex* (pumice), related to *spūma* (foam) as it was thought to form when seafoam hardened

Pumice is a type of rough volcanic glass. It forms when lava froths and quickly cools down as it flies through the air during an explosive eruption. It is full of vesicles – elongated holes left behind by gases escaping from the lava. In fact, pumice contains more air than solid minerals, meaning it is so light that it can float on water. Some volcanic eruptions close to oceans are so vigorous that they produce floating pumice rafts that marine organisms hitch a ride on, carried away by ocean currents. Most pumice is quite pale in colour thanks to a high amount of felsic minerals.

OBSIDIAN

GLASS FORGED IN A VOLCANO

Classification	**Felsic**
Origin	**Volcanic**
Grain Size	**Glassy**
Similar To	**Pitchstone, rhyolite, glass**
Name Origin	**A misspelling of the name Obsius (a Roman who allegedly first found the rock in Ethiopia)**

The high viscosity of rhyolitic magma makes it nearly impossible for crystals to grow. The erupted lava often turns into a volcanic glass with no crystal structure. Obsidian has a typical conchoidal fracture: when it breaks, the complete lack of crystals leads to the development of razor-sharp, curved edges.

Fresh obsidian exposed to the atmosphere is unstable: the randomly arranged silica molecules in glass want to form nice, ordered crystal patterns. Over time, it thus turns to rock in a process called devitrification. The 'snowflakes' (actually crystals of cristobalite) in snowflake obsidian are an example of this change. Because of this process, most obsidian older than 20 million years has been completely devitrified.

1 Obsidian
2 Snowflake obsidian

PITCHSTONE

A WATERY OBSIDIAN

Classification	Intermediate to felsic (can be alkaline)
Origin	Volcanic
Grain Size	Glassy
Similar To	Obsidian, rhyolite
Name Origin	From classical Latin *pix* (sticky)

On touch, pitchstone feels somewhat waxy and looks less shiny than obsidian because it contains more water and crystals in its structure. Some pitchstones even have phenocrysts of alkali feldspar that managed to grow in the magma before being erupted and cooling quickly. Unlike obsidian, it doesn't have a perfectly conchoidal fracture but breaks more randomly due to the higher amount of crystals. But just like obsidian, pitchstone undergoes devitrification: when exposed to air, it loses its resinous shine and turns into a mottled rock. Little orbs of crystals (called spherulites) can grow around microscopic inclusions during the devitrification process.

TRACHYTE

THE ROUGHEST ROCK

Classification	Intermediate (alkaline)
Origin	Volcanic
Grain Size	Fine
Similar To	Phonolite, dacite, rhyolite
Name Origin	From Greek *trākhús* (rough)

Trachyte is the volcanic equivalent of syenite (p. 35). It is made pretty much of only sanidine (a pale alkali feldspar) with some minor occurrences of plagioclase feldspar, amphibole, biotite and pyroxene. The sanidine crystals often grow before the eruption and are found as prominent phenocrysts in the fine, light-grey groundmass. Trachyte is also full of small holes (vesicles) left by escaping gases when the lava cooled down. As the rock weathers, it fractures along these tiny holes, which contributes to the roughness of its surface (hence the name). It is most commonly found in continental rift valleys and at volcanoes behind subduction zones where it erupts alongside rhyolite and andesite.

PHONOLITE

A SINGING ROCK

Classification	Intermediate (alkaline)
Origin	Subvolcanic or volcanic
Grain Size	Fine
Similar To	Trachyte, andesite
Name Origin	From Ancient Greek *phōnē* (voice, sound)

Phonolite is named for the metallic ringing sound it produces when struck. This is likely caused by its small crystals. As the volcanic equivalent of foid syenite (p. 35), it is made of alkali feldspar, nepheline and pyroxene. Phonolitic magma probably forms when alkaline rocks are partially melted in the crust. It tends to produce smaller-scale explosive eruptions, most famously at Mount Erebus in Antarctica, which is also the only volcano with an active phonolitic lava lake inside its crater. The rock can be porphyritic with larger crystals in a fine groundmass (if erupted as lava) or have a uniform medium-grained groundmass if it cools more slowly in volcanic plugs and intrusions just below the surface.

VOLCANICLASTIC

VOLCANICLASTIC ROCKS

From classical Latin *Vulcānus* (the Roman god of fire) + Ancient Greek *klaein* (to break)

Many rocks blur the boundaries between the neat categories we've assigned to them – they are more of a cyclical spectrum. Volcaniclastic rocks are a great example of a whole group that sits at a sweet spot between volcanic igneous and clastic sedimentary rocks. They share their chemical composition with igneous rocks but many of their features resemble those found in sediments.

Just like some volcanic igneous rocks, volcaniclastics are forged in volcanoes during explosive eruptions. Magmas of certain compositions (such as rhyolite, dacite and andesite) tend to explode rather than flow as lava due to their higher viscosity and significant amount of volatile gases (see amygdaloidal rock, p. 22). Volcanic craters can also be plugged by cooled, hardened magma, which exerts even more pressure on the hot, liquid magma underneath. The volatiles within magma expand as they rise towards the surface (exactly like fizzy drinks). At a certain point, the pressure of the plug becomes too much and the sealed magma erupts with a bang, sometimes destroying the volcanic edifice. Explosive eruptions can also happen when magma encounters ice or water and is cooled down rapidly. The sudden change in temperature creates the necessary energy, even if the magma is not full of expanding volatile gases. Because of the immense strength of such eruptions, the magmatic material is torn into smaller particles, called pyroclasts (from Ancient Greek *pûr* – fire + *klaein* – to break). These broken lava fragments fly through the air, fall and blanket the landscape surrounding the volcano.

It is at this point that they enter the realm of clastic sediments. Once pyroclasts settle on the ground, they are collectively called tephra (from Ancient Greek *téphra* – ash). Loose tephra is affected by the same processes as most 'regular' sedimentary grains sourced from weathering rocks (see Sedimentary rocks, p. 68). It can be reworked

and transported by air, water, ice and gravity, mixed with other grains and solidified into rock. Tephra layers are commonly found in glaciers, soil, peat and also as distinct rock units, which can all be used to understand past volcanic eruptions and date various human (pre) historic events. This area of study is called tephrochronology.

Pyroclasts and tephra come in lots of colours, shapes and sizes depending on the composition of the parent magma and how much energy is released during the eruption. They are generally divided into ash, lapilli and volcanic bombs and blocks. Tephra particles tend to fall in a pattern: the largest are found closest to the volcano and their size decreases the further away you go. Many volcanic bombs also gain smooth, aerodynamic shapes as they fly through the air. Because their exterior cools more quickly than the hot interior which keeps expanding, bombs often have cracked, breadcrust-like surfaces.

From our point of view, volcaniclastic rocks are products of the most violent and destructive eruptions. They endanger not only communities living near and on the volcanoes, but also those far away. Volcanic ash is so light that it can be dispersed by winds across the planet and affect everything from weather and transportation to the health of human and non-human species. Yet at the same time, these eruptions create new land and bring nutrients to the Earth's surface. Andosol (from Japanese *an* – dark + *do* – soil) is a type of fertile soil found around many active volcanoes. Important crops – including rice, maize, coffee and tea – are grown in it.

TUFF

AN EXPLOSIVE ROCK

Composition	**Basalt, andesite, dacite, rhyolite, trachyte, phonolite, carbonatite**
Grain Size	**Fine to coarse**
Similar To	**Sandstone, ignimbrite**
Name Origin	**From Latin *tophus* (a loose, porous rock)**

Tuff is made of tephra that fell on the ground and solidified to form a tough rock. While most tuffs are mainly made of ash, they can contain other components, including lapilli (lapilli tuff), fragments of rocks torn from the ground by the ascending magma (lithic tuff), crystals of various minerals (crystal tuff) or shards of volcanic glass (vitric tuff). Some deposits of tuff have distinct layers – similar to beds in clastic sedimentary rocks – caused by multiple volcanic eruptions happening one after the other. If the falling tephra gets mixed with other sediments as it falls on the ground, it can create a rock called tuffite. Depending on its composition and age, tuff can have a range of colours, from dark red to light green, grey and pale yellow.

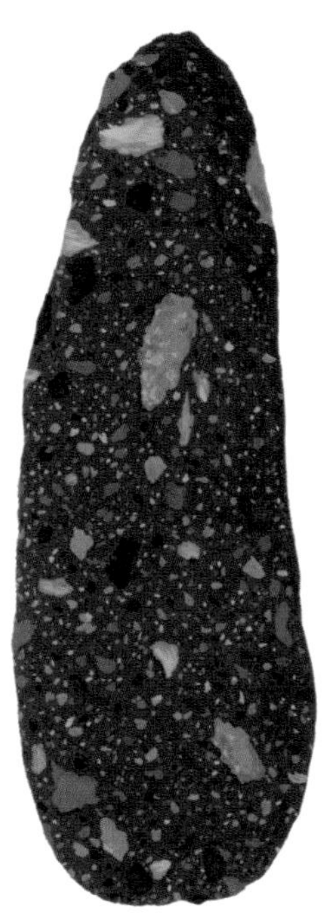

1 Layered tuff
2 Lapilli tuff
3 Lithic tuff

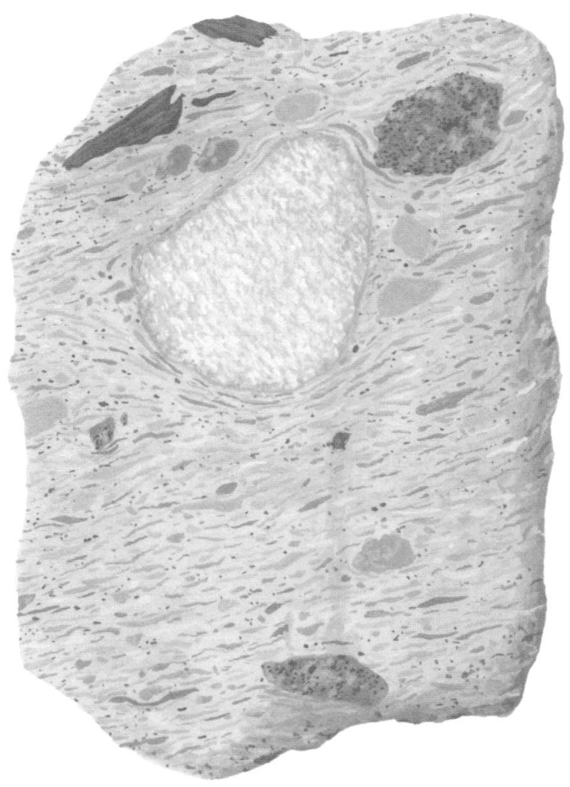

IGNIMBRITE

THE MOST DESTRUCTIVE ROCK

Composition	**Andesite, dacite, rhyolite, trachyte, phonolite**
Grain Size	**Fine to coarse**
Similar To	**Tuff**
Name Origin	**From Latin *ignis* (fire) + *imber* (rainshower, stormcloud)**

Ignimbrite (also known as welded tuff) is the deposit left by pyroclastic density currents (also called *nuée ardente*, French for 'burning cloud', or simply PDCs). These are turbulent avalanches of hot tephra, lava and volcanic gases that move at incredibly high speeds (up to hundreds of kilometres per hour) down the flanks of explosive volcanoes. Pyroclastic density currents are responsible for some of the most destructive eruptions: the incandescent ash cloud blankets the landscape and burns everything (and everyone) in its path. Eventually, it settles down but because it's still hot (over 600°C or 1,110°F), the ash welds itself into a solid rock. Some of the larger glassy fragments and pumice get squished by the weight of the material above into elongated features called fiamme.

←
Mount Vesuvius
Italy

The infamous eruption of Mount Vesuvius in 79 CE hurled pyroclastic density currents at the settlements of Pompeii, Herculaneum, Stabiae and Oplontis. Their buildings and inhabitants were buried under layers of hot ash, which eventually hardened into ignimbrite. Letters from Pliny the Younger, which describe this eruption, are the first written record we have of volcanic activity.

SCORIA

A ROCK FULL OF HOLES

Composition	Basalt, andesite, trachyte, phonolite
Grain Size	Fine
Similar To	Pumice, tuff
Name Origin	From Greek *skór* (dung)

Just like pumice (p. 49), scoria is full of small holes left by gas bubbles that escaped the lava as it erupted. However, scoria contains denser, iron-rich minerals (such as olivine and pyroxene), so it doesn't float on water like pumice.
It normally doesn't fly very far from the eruption site. Instead, it accumulates around the crater to form a type of volcano known as a cinder cone. Once exposed to air, scoria often rusts: it oxidizes from its initial grey colour into shades of red and brown thanks to the high amount of iron.

AGGLOMERATE

A ROCK FULL OF VOLCANIC BOMBS

Composition	Basalt, andesite, dacite, rhyolite, trachyte, phonolite
Grain Size	Medium to coarse
Similar To	Conglomerate, lithic tuff
Name Origin	From Latin *ad* (together) + *glomerāre* (to form into a ball)

Agglomerate is most often found filling the craters of volcanoes. It forms during powerful eruptions that throw fragments of rocks and volcanic bombs into the air. Most of this material falls right back into the vent it came from, effectively sealing it. Over time it solidifies to form an agglomerate. Given its explosive nature, the structure of this rock tends to be highly chaotic, with rounded bombs of many different sizes randomly distributed in the surrounding matrix of ash or lava.

1 Explosion breccia
2 Intrusive breccia

IGNEOUS BRECCIA

FRACTURED LAVA

Composition	Basalt, andesite, dacite, rhyolite, trachyte, phonolite, carbonatite
Grain Size	Medium to coarse
Similar To	Sedimentary breccia
Name Origin	Borrowed from Italian *breccia* (gravel, rubbish of broken walls)

Igneous breccia (pronounced 'bre-tcha') looks similar to its sedimentary counterpart (p. 83): it contains angular fragments in a matrix. Intrusive breccia forms when a fresh batch of magma is intruded into already cool and solid igneous rock, fracturing and embedding it in the fresh magma in the process. During highly explosive events, the force of the eruption can tear away bits of the bedrock and the tephra solidifies around them, creating explosion breccia. Flow breccia (or autobreccia) forms when bits of cooled lava break apart and are encased in the advancing lava flow. Finally, pyroclastic breccia contains fragments of volcanic bombs.

HYALOCLASTITE

LAVA MEETS WATER

Composition	Basalt, andesite, dacite, rhyolite, trachyte, phonolite
Grain Size	Fine
Similar To	Tuff
Name Origin	From Greek *ýalos* (glass) + *klaein* (to break)

When hot lava erupts beneath a cold glacier or underwater, it quickly becomes explosive. The temperature shock rapidly shatters the lava into small glassy fragments. The resulting rock is a type of tuff called hyaloclastite, which is often found alongside various types of volcanic glass (like tachylyte, p. 43). The continued interaction with water alters it, creating a buff-coloured palagonite with embedded pieces of dark glass and crystallized lava. Hyaloclastite is a typical rock found at underwater seamounts, inbetween lava pillows and on subglacial volcanoes called tuyas.

SEDIMENTARY

SEDIMENTARY ROCKS

From Latin *sedēre* (to settle, to sit)

Three-quarters of the Earth's surface is blanketed by sedimentary rocks, which are being covered by sediments forming today. Despite their name, most sedimentary rocks are full of action and movement.

Chemical sedimentary rocks form when fluids (mainly water) contain too many dissolved minerals, so they begin to precipitate out. This process most often happens in the oceans, but can form rocks on land.

Biochemical (or organic) sedimentary rocks are formed by living organisms. They can be divided into rocks chemically precipitated from shells and skeletons of marine organisms, and those rich in carbon.

Clastic sedimentary rocks are the most common, and begin as grains (also called clasts) weathered from a pre-existing rock. The mineral and rock grains are carried by air, water, ice and gravity away from their source rock and deposited elsewhere. Any breaks in deposition create distinct layers called beds (if they are thicker than 1 centimetre or 0.4 inches) or laminations (if they are thinner than that). The layers are then buried by more sediment and compacted. Finally, the grains are cemented together and turned into solid rock in a process called diagenesis.

1. Mud
Less than 0.004mm

2. Silt
0.004 to 0.06mm

3. Sand
0.06 to 2mm

4. Granule
2 to 4mm

5. Pebble
4 to 64mm

6. Cobble
64 to 256mm

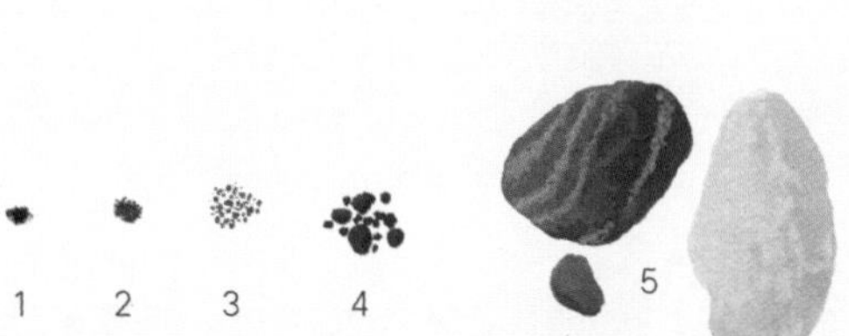

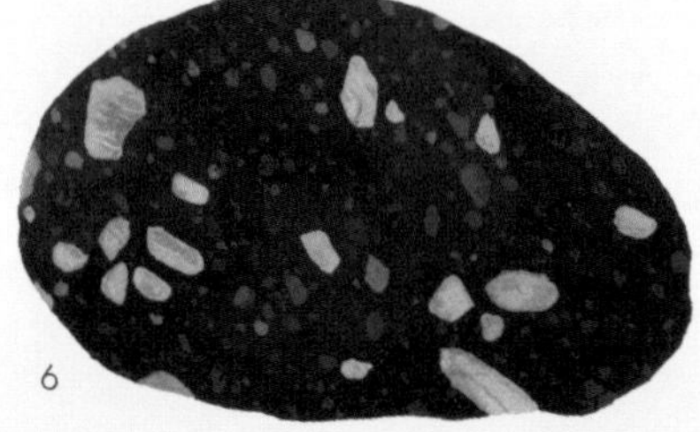

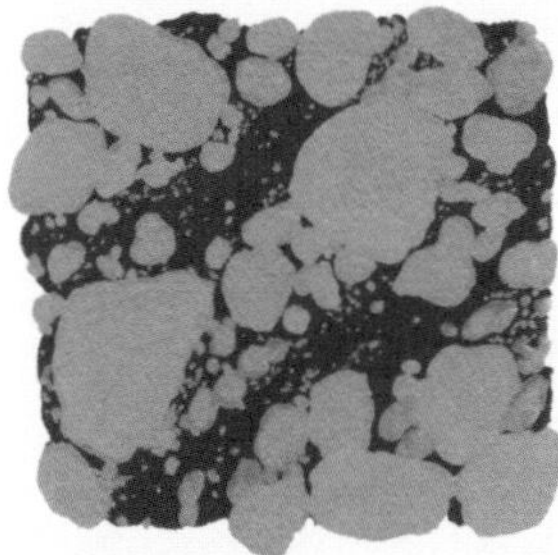

↑
Poorly sorted
A jumble of different grain sizes

↗
Well sorted
All grains are a similar size

↗↗
Porosity
Empty spaces between grains

The grain size of clastic sedimentary rocks tells us how far the original sediment moved. Each grain can be classified according to its size, in ascending order as mud, silt, sand, granule, pebble, cobble or boulder. Bigger grains need a lot of energy to be moved, so they tend to be found closer to the source rock. Small grains can be carried far away as they require much less energetic environments to get moving. Some rocks have larger clasts surrounded by a finer matrix.

If grains look smooth and spherical, they probably moved quite far from their source rock and became rounded by constant collisions with other grains. Jaggy and angular grains most likely travelled only a short distance before being deposited.

Sorting refers to the distribution of grains within a rock. Well sorted rocks form over longer periods of time in environments with more or less constant energy (think rivers, deserts and beaches). Poorly sorted rocks are most often formed during a single chaotic event (like flood surges and storms).

Sometimes, the spaces between grains remain empty: this makes the rock porous. The pores can hold groundwater, oil or natural gas.

Sedimentary rocks that contain fossils are also the archive of the evolution of life on Earth. It's a very patchy archive, though, because not all life gets fossilized. Ideally, a future fossil needs to be rapidly buried by sediment to protect it from scavengers and erosion. The soft tissue breaks down, while the more resistant bones or exoskeleton are left. Mineral-rich fluids then dissolve and replace the hard parts with other minerals (usually calcium carbonate, iron minerals or silica) or the empty spaces get filled with sediment, creating a cast.

Trace fossils can tell us how an organism lived. Also known as ichnofossils, these include footprints and traces of movement, burrows, feeding traces and coprolites (fossilized poo) that were buried shortly after they were made.

PALAEOSOL

FOSSILIZED SOIL

Classification	Clastic, chemical
Environment	Continent
Grain Size	Very fine to fine
Similar To	Claystone, siltstone
Name Origin	From Ancient Greek *palaiós* (old, ancient) + Latin *solum* (ground, soil)

Soils are a relatively new addition to the geological record. It wasn't until after plants learnt how to cooperate with fungi to grow on dry land (around 470 million years ago) that the Earth's bare rocks began to be broken down and mixed with organic matter to form the first soils. Their colour reveals a lot about the local climate at the time of their formation, including how much it rained and how hot or cold it was.

Red and yellow soils owe their colour to oxidized iron minerals, which tend to form in

well-drained conditions. Laterite is an example found on top of basalt in wet and humid tropical areas. Iron-rich palaeosols are the sources of red and yellow ochres, which were used as earth pigments in prehistoric cave paintings.

Grey, green and blue soils form when microbes reduce iron in the absence of oxygen. The waterlogged gleysols and seasonally frozen cryosols of tundra are typical examples.

Dark-brown to black soils are full of organic matter. They range from the fertile mollisols to waterlogged histosols, like peat. The mineral vivianite (blue ochre) grows in bogs when phosphorus is released from decaying organisms.

Pale-coloured soils are usually abundant in clay minerals, while white palaeosols like calcrete (a type of tufa, p. 92) contain calcium carbonate.

1 Yellow ochre 2 Cryosol 3 Red ochre
4 Calcrete 5 Histosol with vivianite

CLAYSTONE

A ROCK CHANGED BY WATER

Classification	Clastic
Environment	Ocean, lake, continent
Grain Size	Very fine
Similar To	Palaeosol, mudstone
Name Origin	From Old English *clæg* (sticky earth, clay)

Clay minerals are found in most clastic sedimentary rocks as the glue that holds the grains together. If they make up more than half of the grains, the rock is called claystone. Often found unconsolidated, clays form in two main ways: during weathering and when low-temperature hydrothermal fluids alter rocks. These processes introduce water into the chemical structure of pre-existing minerals and change them into clay minerals, such as talc, vermiculite, kaolinite, illite, smectite, chlorite or muscovite. The weathering of volcanic tuff makes a claystone called bentonite. Kaolin is formed when feldspars in granites and gneisses are weathered and changed into kaolinite. Certain clays expand and become plastic when wet, making them perfect for ceramics (see pottery, p. 140).

1 Kaolin
2 Bentonite

UMBER

A PIGMENT FROM THE DEEP SEA

Classification	Clastic, chemical
Environment	Deep ocean (hydrothermal vents)
Grain Size	Very fine
Similar To	Claystone
Name Origin	Either from Latin *umbra* (shadow) or after Umbria, a region in Italy where it was mined

Near mid-ocean ridges, black smokers powered by upwelling magma spew out superheated water mixed with various metallic compounds (mostly pyrite, an iron sulphide responsible for their black colour). As soon as the hot fluid comes into contact with the cold seawater, some of the minerals it contains precipitate near the vent to build the typical chimney-like structures. Most of it gets carried away into the deep ocean, though, where manganese and iron oxides precipitate and settle on the seafloor as a layer of fine clayey sediment called umber. Umber is known as an earth pigment – it was one of the first pigments to be used by humans, alongside sienna (which has less manganese) and ochre (rich in iron).

MUDSTONE

THE MOST COMMON SEDIMENTARY ROCK

Classification	Clastic
Environment	Deep ocean, deep lake, lagoon, delta
Grain Size	Very fine
Similar To	Claystone, shale
Name Origin	From Middle Low German *mudde* (soft earth)

Mudstone is made of grains so light that they can be carried even by the slowest currents into the deep ocean. Its colour is based on the amount of organic matter the muddy sediment contains and the environment it forms in. Black is caused by carbon derived from organisms, while grey to olive green means the organic content is low. The black to olive green scale also indicates the sediment deposited in an environment with little oxygen (like the bottom of an ocean), while redder muds form in shallower, more oxygenated waters. It can contain fossils of various marine organisms or the perfectly cubic crystals of pyrite (aka fool's gold).

SHALE

MUDSTONE, BUT WITH LAYERS

Classification	Clastic
Environment	Deep ocean, deep lake, lagoon, delta
Grain Size	Very fine
Similar To	Mudstone
Name Origin	From Old English *scealu* (a thing that divides)

Shale is just like mudstone in terms of grain size (and forms in the same environments) but has an extra trick up its sleeve: it is fissile. When struck, shale breaks into thin layers along its bedding planes. It is the orientation of the platy clay minerals that causes fissility. As the layers of mud are pushed down and squished by the fresher sediment above, they align in the same direction. This creates tiny lines of weakness between the individual layers. Shale can contain fossils of animals that fell into the sediment when they died. The specimen below has needle-like graptolites, a type of extinct colonial plankton. Some of this organic material may turn into kerogen (the first step in the formation of natural oil and gas), creating oil shale.

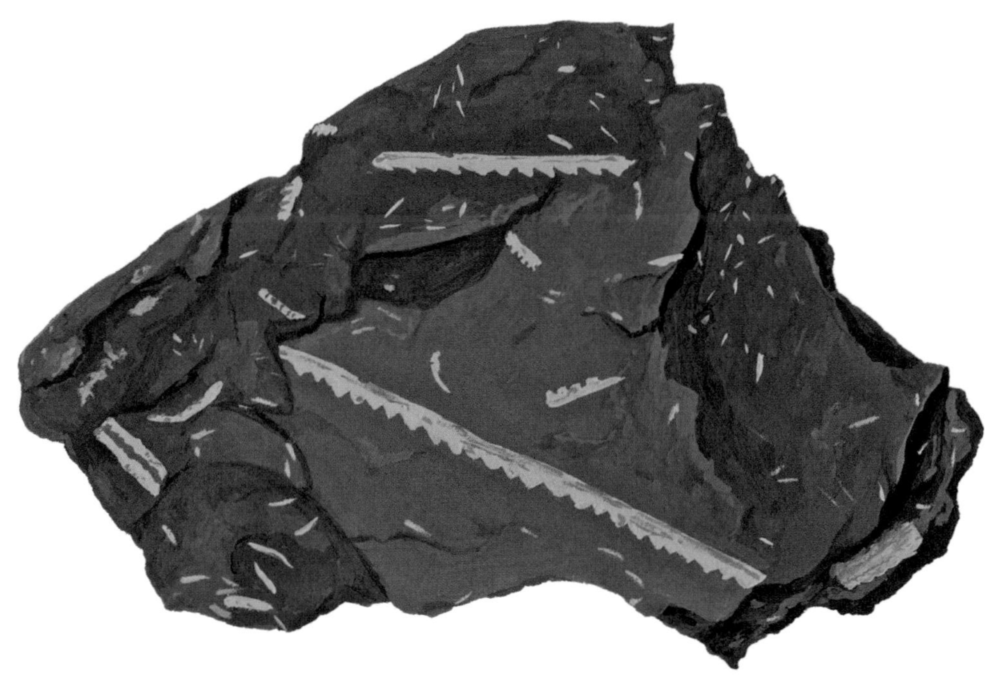

MARL

PART MUDSTONE, PART LIMESTONE

Classification	Clastic
Environment	Freshwater lake, shallow sea, saline basin
Grain Size	Very fine
Similar To	Mudstone, limestone
Name Origin	From Gaulish Celtic *marga* (earth)

Marl is a type of mudstone with a high amount of calcium carbonate. It is essentially a rock transitional between mudstone and limestone. White marl most often forms in alkaline freshwater lakes when quartz silt is washed from land into rivers and then carried into the lakes. The calcium is supplied by dead algae (such as the fittingly named stonewort), which create calcite around their cells. Marl can also be found in shallow, oxygen-poor seas. Marine marl has a distinct grey to tea-green colour caused by the mineral glauconite, the source of pigment terre verte (green earth). Another place marl forms is in saline basins that periodically dry up, where it can be found between beds of gypsum and rock salt.

SILTSTONE & LOESS

ROCKS OF GLACIERS AND DELTAS

Classification	Clastic
Environment	Ocean, lake, river, glacier
Grain Size	Very fine
Similar To	Mudstone, claystone
Name Origin	Silt from Old High German *sulza* (brine, salt marsh); loess from German *lösch* (loose)

Silt-sized grains are larger than mud but still light enough to be carried by low-energy flows. They most often settle at the bottom of lakes, in river deltas and on the ocean floor where they are deposited by submarine mass flows. Silt can also be created by glaciers as they pluck and grind the bedrock they flow over. Before it hardens into siltstone, this kind of sediment is called rock flour, boulder clay or diamicton. Generally speaking, siltstone contains more quartz than mudstone, is often greyish or yellowish and lacks fissility.

Buff-coloured loess accumulates on the frigid fringes of continental glaciers, where fast winds shift silt-sized angular grains of quartz, feldspar and mica. It is usually cemented by calcium carbonate.

1

2

1 Siltstone

2 Loess

←
Lake Louise
(Horâ Juthin Îmne)
Canada

When meltwater with suspended rock flour flows into lakes and rivers, it can change their colour to a cloudy turquoise or teal (the exact hue depends on how much sediment is in the water). This phenomenon occurs when light is scattered by the small silt grains. This accounts for the vibrant shade of blue you can see at Lake Louise and other glacial lakes. Over time, the soft sediment solidifies into siltstone.

SANDSTONE

THE SECOND MOST COMMON SEDIMENTARY ROCK

Classification	Clastic
Environment	Sea, river, desert
Grain Size	Fine to medium
Similar To	Siltstone
Name Origin	From Old English *sand* (sand)

Up to a quarter of all sedimentary rocks on Earth are sandstone. It is most often made of quartz, but feldspar, mica or rock fragments are also commonly present. We can distinguish between several types of sandstone.

Fluvial sandstone (from Latin *fluvius* – river) forms in rivers and river deltas. Its grains are usually more angular, less well sorted and tend to be yellowish to greyish. It can have distinct cross-bedding: tilted layers within its beds that represent the individual current ripples.

Aeolian sandstone (from Ancient Greek *Aeolus* – a mythical ruler of the winds) is formed in deserts. It is made of rounded, well sorted

1 Orthoquartzite 2 Fluvial sandstone
3 Greensand 4 Aeolian sandstone
5 Arkose 6 Greywacke

grains of quartz, smoothed as they are blown around by fast winds. Sand dunes are often preserved as large-scale cross-bedding. The characteristic orange-red colour is a result of exposure to air: iron-rich haematite in the sand coats the individual grains in a thin rusty layer.

The whitish to pinkish orthoquartzite is a real quartz extravaganza – more than 95% of its grains are pure quartz. They are cemented together by even more quartz, which makes the rock extremely hard and resistant to erosion. It normally forms in rivers or shallow seas.

Arkose is solidified grus, a type of sand made of feldspar, quartz and mica, which forms from the weathering of granite and gneiss. The high feldspar content (at least 25%) gives arkose its distinct dull pink to brown colour.

Greywacke has a grey colour due to its high content of mud. It forms in deep ocean environments as a result of fast-moving underwater avalanches called turbidity currents.

Greensand (or glauconitic sandstone) is created in shallow seas with low amounts of oxygen. There, the sand grains are mixed with excrement of marine organisms which give the sediment its distinct pale green colour. It can contain various marine fossils (this sample has bivalves).

CONGLOMERATE

A SOLIDIFIED RIVER

Classification	Clastic
Environment	River
Grain Size	Medium to coarse
Similar To	Diamictite, tillite
Name Origin	From Latin *con* (together) + *glomerāre* (to form into a ball)

Conglomerate (also known by its more poetic name, puddingstone) is a common rock made of large rounded clasts: pebbles, cobbles and boulders. Most conglomerates are formed in rivers, which have enough energy in their flow to transport such large grains. Depending on what kind of bedrock the ancient river flowed over, the grains in a conglomerate can all be the same rock and/or mineral (we call such conglomerates monomict), or a mix of various types (polymict). The size and rounding of the grains indicates if they were sourced from far away (more rounded and smaller) or nearby (less rounded and larger).

SEDIMENTARY BRECCIA

A BROKEN ROCK

Classification	Clastic
Environment	Mountains
Grain Size	Medium to coarse
Similar To	Igneous breccia, impact breccia, diamictite
Name Origin	Borrowed from Italian *breccia* (gravel, rubbish of broken walls)

Before sedimentary breccia becomes sedimentary breccia, it is scree accumulated at the base of a cliff, a landslide, a rockslide or a deposit left by a mudflow. The composition of its clasts depends on the original rock(s) that the fragments came from. Much like conglomerate, sedimentary breccia can be monomict (made of one type of rock or mineral) or polymict (a mixture of different rocks and minerals). The key feature that differentiates it from a conglomerate is the sharp edges of its clasts. It tells us that they didn't travel very far from their source rock and were not smoothed by the flow of water. The clasts can be surrounded by silt, sand and cement in various proportions.

DIAMICTITE

A TOTAL MESS

Classification	Clastic
Environment	Deglaciated continent, mountains, deep sea
Grain Size	Fine to very coarse
Similar To	Tillite, sedimentary breccia, conglomerate
Name Origin	From Greek *dia* (through) + *miktós* (mixed)

Diamictite is a descriptive term used for any poorly sorted rock consisting of a mix of large and small fragments in a finer matrix. It is often associated with deglaciated landscapes. Diamictite can be formed directly inside glaciers as they grind the bedrock (in that case it's called tillite). It also forms during calving, when large blocks of ice break from the edge of a glacier and float into the open ocean as icebergs. They can raft rocks far away from land. As the iceberg melts, it plunges its rocky passengers into the soft sea sediment to create dropstones (pictured). Not every diamictite is glacial in origin, though. Other ways to form it include volcanic mudflows (also known as lahars – a mix of erupted ash and water), debris flows (powerful movements of water and sediment in mountains following intense rain) and deep-sea avalanches (called turbidites).

TILLITE

A WITNESS OF MELTING GLACIERS

Classification	**Clastic**
Environment	**Deglaciated continent**
Grain Size	**Fine to very coarse**
Similar To	**Diamictite, sedimentary breccia, conglomerate**
Name Origin	**From Old English *tilian* (to cultivate, to tend)**

Glaciers are dirty. As they flow, they pluck bits of rock from the bedrock beneath and rocks also fall on top of them from the surrounding mountains. The constant movement of the ice slowly breaks up and grinds most of the rocks into a silty sediment. When the glacier melts, all the fine sediment and any bigger rocks are dumped on the ground. As time passes, this poorly sorted material can solidify into tillite (a type of diamictite). Sometimes, only the largest rounded boulders get preserved in the landscape – these are known as glacial erratics. It is possible to figure out where the ancient ice sheet flowed from by identifying the rock fragments the tillite contains, particularly if they don't match the local bedrock.

BEACHROCK

THE FASTEST-FORMING SEDIMENTARY ROCK

Classification	Clastic
Environment	Coast
Grain Size	Fine to coarse
Similar To	Sedimentary breccia, limestone
Name Origin	Probably from Old English *bece* (stream)

Beachrock is (unsurprisingly) found as a thin blanket covering older rock outcrops on seashores. It forms in the intertidal zone, where salty seawater mixes with freshwater coming from the land. This interaction allows for calcium carbonate to deposit and turn to solid rock relatively quickly, in the range of mere tens to hundreds of years. The carbonate cement binds together any material found on the beach, such as rocks, shells, corals and even archaeological human artefacts. Because beachrock forms so rapidly, it can be used to study how local sea levels changed in our most recent history.

LIMESTONE

A ROCK IS BORN

Classification	Biochemical
Environment	Shallow sea
Grain Size	Very fine to fine
Similar To	Dolomite
Name Origin	From Old English *lim* (mud)

Marine organisms like corals, brachiopods, crinoids (sea lilies), bivalves, oysters, plankton and algae use calcium carbonate in their bodies for protection (in shells) or support (in skeletons). When they die, their bodies turn to rock. Algae in particular are some of the most vigorous producers of micrite – a yellowish to greyish carbonate mud that forms the bulk of most limestones. Dark limestones usually contain lots of organic matter or mud. Fossilized shells and skeletons of other marine organisms are often present as well, offering a snapshot of life in an ancient sea. The rock easily dissolves in acidic waters; a lot of limestone landscapes are full of cave systems and sinkholes known as karst (see travertine, p. 93). This chemical reaction works on hand specimens too. If you drop vinegar or lemon juice on the rock, the calcite will fizz as it turns into carbon dioxide.

1

2

1 Coral limestone

2 Crinoidal limestone

←

The Great Barrier Reef
Australia

The world's largest coral reef system hugs the northeast coast of Australia. Today's corals grow on top of an older reef platform – the building blocks of future limestone. However, the reef is under threat, with the rising ocean temperatures causing coral bleaching. The colourful symbiotic algae get expelled from the coral, revealing the white calcium carbonate of the coral skeleton.

DOLOMITE

LIMESTONE'S CONFUSING COUSIN

Classification	Chemical
Environment	Shallow sea
Grain Size	Very fine to fine
Similar To	Limestone
Name Origin	After the French mineralogist Déodat de Dolomieu (1750–1801) who first described it

Dolomite is similar in composition to limestone but instead of calcite, it is made of a carbonate mineral called... also dolomite (confusing, I know). Because of this, dolomite (the rock) is sometimes called dolostone. Dolomite (the mineral) probably forms when buried lime sediments are periodically dissolved and reprecipitated in water with changing salinity (perhaps also aided by the growth of bacteria). However, nobody is quite sure how exactly this process happens and this conundrum is known as 'the dolomite problem'. What we do know is that the process tends to destroy any features (fossils included) and the resulting rock is a darker earthy grey with a sugary texture. It also won't fizz as much as limestone in acid because dolomite (the mineral) contains a mix of magnesium and calcium.

OOLITE

EGG ROCK

Classification	Chemical
Environment	Shallow sea
Grain Size	Fine
Similar To	Pisolite
Name Origin	From Ancient Greek *ōión* (egg)

Oolite is made of small (below 2 millimetres), shiny, egg-like spheres called ooids. To form, ooids need three things: a warm marine environment that is not too deep; water saturated with calcium carbonate; and a lot of wave action. At first, a thin layer of calcium carbonate forms around a nucleus (usually a sand grain or a shell fragment). As the ooid is rolled around by the waves, it gradually accumulates more and more concentric layers. After a deposit of ooids is covered by more sediment, the tiny spheres get cemented together and turn into an oolite.

PISOLITE

A GROWN-UP OOLITE

Classification	Chemical
Environment	Continent, shallow sea
Grain Size	Coarse
Similar To	Oolite
Name Origin	From Ancient Greek *píson* (pea)

Pisolite has spherical grains larger than 2 millimetres that are called pisoids (or pisoliths). These concentric grains form in a similar way to ooids, when carbonate mud coats a nucleus (such as a mineral grain) layer by layer, causing it to gradually increase in size. Pisolite can be found in shallow seas, the vadose zone (the area between the soil surface and the top of the water table), terrestrial caves and calcrete soils (p. 71). In tropical environments with alternating wet and dry seasons, pisolite is often stained red by iron and its calcium carbonate is replaced by clays and bauxite (an aluminium ore).

TUFA

LEFTOVER CALCIUM

Classification	Chemical
Environment	Hot spring, river, lake, limestone cave
Grain Size	Fine
Similar To	Travertine
Name Origin	From Latin *tophus* (a loose, porous rock)

The white to yellowish tufa is a soft and porous rock formed when calcite precipitates out of water rich in carbonates, normally around limestone caves, hot springs, in places where a river dried up or in deserts. It is sometimes found as a hard, crusty layer alongside ancient soil sediments – in that case it's known as calcrete (p. 71). Various algae, mosses and bacteria like to live near or directly on tufa, which results in them getting encrusted in the calcium carbonate and preserved as part of the rock. Not to be confused with the similarly sounding but altogether differently forming tuff (p. 58).

TRAVERTINE

LIMESTONE GROWN IN CAVES

Classification	Chemical
Environment	Hot spring, limestone cave
Grain Size	Fine
Similar To	Tufa, calthemite
Name Origin	From Latin *tiburtinus* (relating to the ancient town of Tibur in Italy)

Travertine forms in karst landscapes when limestone (and sometimes dolomite) is dissolved by weak acids in rainwater. It then flows to underground caves, drips and precipitates as speleothems: stalactites (if they hang from the ceiling), stalagmites (if they grow from the floor) or columns (if stalactites and stalagmites connect). The other environment where travertine is often found is hot mineral springs. There, calcium carbonate precipitates out of water as it dries up, forming mounds, dams and cascade deposits. It is more compact than tufa, with rounded surfaces, and often has distinct banded layers formed by the repeat precipitation.

CHALK

THE ROCK THAT NAMED A GEOLOGICAL PERIOD

Classification	Biochemical
Environment	Deep ocean
Grain Size	Very fine
Similar To	Diatomite
Name Origin	From Latin *calx* (lime, small stone)

Chalk is a variety of limestone made of microscopic plankton, such as coccolithophores (coccoliths for short) and foraminifera. These organisms use calcium carbonate to build intricate shells around their bodies in order to protect themselves from predators. When they die, their empty shells rain onto the seafloor in their billions. There, they get buried and compacted, and over time harden into a solid yet porous rock. Chalk can contain fossils of larger organisms, too. The needle-like spicules (structural elements) of sea sponges are found most often; these animals use silica to build their bodies, which often turns into a layer of flint. Chalk is very common in deposits from the Cretaceous (145 to 66 million years ago) and this whole geological period was named after it (*crēta* is the Latin word for chalk).

CHERT & FLINT

ROCKS THAT DEFINED THE STONE AGE

Classification	Biochemical
Environment	Deep ocean
Grain Size	Very fine
Similar To	Radiolarite
Name Origin	Chert is probably a local English word; flint from Old English *flint* (a hard rock which gives off sparks when struck)

Some marine organisms (like radiolarians, diatoms and sea sponges) use silica to build mineral skeletons for support or protection. The formation of chert begins when they die and their bodies fall as siliceous ooze into the sediment on the ocean floor. Their shells get dissolved in seawater and turn into a silica-rich goo. This material undergoes various chemical changes and hardens into chert. Clays and mud can also be present, in which case the resulting rock is often known as porcellanite.

Flint is a dark type of chert typically found as layers or nodules within chalk. It often has a light-coloured powdery rind caused by weathering. Flint breaks with a conchoidal fracture that creates sharp curved edges, which made it a popular material for tools during the Stone Age.

1 Chert
2 Flint

DIATOMITE

A ROCK MADE OF PHYTOPLANKTON

Classification	Biochemical
Environment	Deep ocean, deep lake
Grain Size	Fine
Similar To	Chalk
Name Origin	From Greek *diátomos* (cut in half)

Diatomite is a whitish rock made of diatoms – microscopic single-celled algae. They are a part of phytoplankton, a collection of mainly algae and cyanobacteria (a group of bacteria that obtains food via photosynthesis). Phytoplankton make up the base of ocean and freshwater food webs and are one of the most important producers of oxygen on the planet. When the conditions are right, they can multiply fast, covering the surface of the water in an algal bloom. After a life of photosynthesizing, billions of diatoms die together and fall to the bottom of an ocean or a lake as siliceous ooze. As their bodies accumulate in the abyss and get buried by sediment, the silica in their shells slowly turns into opal. Diatomite crumbles easily into a fine powder called diatomaceous earth.

RADIOLARITE

A ROCK MADE OF ZOOPLANKTON

Classification	Biochemical
Environment	Deep ocean
Grain Size	Fine
Similar To	Chert
Name Origin	From Latin *radiolus* (a little radiating spike)

Radiolarite (or radiolarian chert) forms similarly to chert when siliceous ooze settles at the bottom of an ocean. But in this case, the ooze is made exclusively of dead radiolarians, a type of microscopic zooplankton. Alongside radiolarians, zooplankton is made of other single-celled organisms (called protozoans), small crustaceans and larvae. These graze on the even smaller phytoplankton and are, in turn, eaten by larger creatures. Once they fall to the ocean floor, their tiny shells dissolve, the silica in them transforms into opal and then into microcrystalline quartz. The resulting rock is usually dark brown or deep orange-red and can also contain other minerals that settle on the seafloor alongside the ooze, such as various clays.

BANDED IRON FORMATION

A WITNESS OF ATMOSPHERIC CHANGE

Classification	Chemical
Environment	Shallow sea
Grain Size	Fine
Similar To	Chert
Name Origin	Iron likely borrowed from Celtic *īsarno-* (iron ore) during the Iron Age

Banded iron formation (BIF for short) truly lives up to its name: it's a rock formation (check) with bands (check) and is rich in iron ores such as haematite and magnetite (check). The silvery-to-black layers of ore alternate with red chert, giving this rock its distinct look. But there is a catch: it can't form anymore. Most BIFs formed about 2.4 to 1.8 billion years ago during a time when there was no oxygen in the atmosphere or the oceans. Instead, water was full of dissolved iron emitted from deep-sea hydrothermal vents. But then, cyanobacteria learnt how to turn carbon dioxide into sugars through photosynthesis. Oxygen was just a byproduct that filled the oceans during what is known as the Great Oxidation Event. The free-floating iron suddenly found itself bonded to this new, abundant element and began to accumulate on the seafloor. Once all free iron was used up, BIFs stopped forming and oxygen could start filling the atmosphere.

GEYSERITE

A ROCK OF HOT SPRINGS

Classification	Chemical
Environment	Geyser, hot spring
Grain Size	Very fine
Similar To	Chert
Name Origin	From Old Norse *gøysa* (to gush)

Also known as siliceous sinter, geyserite is a type of chert formed in areas where felsic magma rich in silica is close to the surface. Its warmth heats up deep groundwater that then rises towards the surface through cracks in the bedrock, dissolving silica from the rocks along the way. Once the superheated water erupts on the surface and cools down, the silica gets deposited around the vent. Any microbes or plants living near the active geyser may be encrusted in the rock, too. Rhynie chert from Scotland (pictured) is the oldest preserved terrestrial ecosystem full of early plants, fungi, lichens and arthropods, which were fossilized by geysers around 410 million years ago.

AMBER

FOSSILIZED TREE RESIN

Classification	Biochemical
Environment	Continent (forest)
Grain Size	No grains
Similar To	Resin
Name Origin	From Arabic *'anbar* (ambergris, a substance produced in the intestines of sperm whales)

Not just any hardened tree resin can be called amber: its molecules have to be polymerized – arranged in chains. That can only happen when resin is buried in sediment and exposed to increased pressure and temperature. The oldest amber comes from the Carboniferous Period (359 to 299 million years ago) but it is much more common from the Cretaceous Period (145 to 66 million years ago) when plants first started flowering. It can vary in colour from yellow-orange to pale yellow, deep brown and green. It can be transparent (meaning you can see right through it) or translucent (only light gets through). Amber often contains fossilized plants, fungi, insects, arachnids or even small vertebrates that got accidentally trapped in the sticky resin.

COAL

FOSSILIZED SUNLIGHT

Classification	Biochemical
Environment	Continent (bog)
Grain Size	Variable
Similar To	Black mudstone
Name Origin	From Old English *col* (burning charcoal)

1

Most coal comes from fossilized plants that grew in the Carboniferous Period (359 to 299 million years ago). Back then, the tropics were covered in sprawling swampy forests. Tree-like relatives of clubmosses and horsetails happily photosynthesized, releasing vast amounts of oxygen into the atmosphere and sequestering carbon dioxide. They grew fast and died young, toppling into the bog below when they became too tall. The waterlogged, oxygen-poor conditions prevented fungi and microorganisms from decomposing the wood and releasing its carbon back into the atmosphere. Instead, the carbon remained intact as the plant turned to peat.

2

Over time, pressure and temperature gradually increased the proportion of carbon, turning peat into lignite (brown coal), then sub-bituminous coal, bituminous coal and, finally, anthracite (black coal). This process is called coalification. It's the reason coal burning contributes so much to global warming. We are putting the atmosphere off balance by filling it with carbon that was locked away in the rocks for 300 million years.

3

1 Peat
2 Bituminous coal
3 Anthracite

←
The Flow Country
Scotland

Coal starts its life as peat, a type of soil that forms in wetland areas such as bogs and fens. The largest blanket bog in Europe is the Flow Country in the far north of Scotland. Its deep layers of peat have been forming for around 9,000 years at a rate of 1 millimetre per year. Peatlands soak up water like a sponge, thus acting as a natural protection against flooding, store vast amounts of carbon, and play a crucial role in mitigating the effects of the climate crisis. Unfortunately, many are being degraded through commercial cutting, draining and burning.

METAMORPHIC

METAMORPHIC ROCKS

From Ancient Greek *metá* (to change) + *morphé* (having a shape)

It is easier to imagine how sedimentary, igneous and volcaniclastic rocks form – we witness some of the processes that create them every day. However, most rocks will be changed by metamorphism at some point during their existence. Understanding metamorphic processes is a bit trickier because they normally happen deep inside the Earth's crust in extreme environments. The further underground rocks are, the more material lies above them (increasing the pressure they are under) and the closer to the hot core they are (increasing the temperature). During metamorphism, the original rock (called protolith or parent rock) recrystallizes: its features and minerals transform due to higher pressure and/or temperature. There are a few ways in which that can happen.

Contact metamorphism happens locally when magma intrudes into bedrock. The rocks around the intrusion are slowly heated and the closer they are, the more metamorphosed they get. This can occur on both small (a few millimetres around dykes and sills) and large scales (hundreds of metres around batholiths).

Regional metamorphism affects large areas. It is typical in orogenic (mountain-building) events during the collision of two tectonic plates. Rocks are metamorphosed at different depths under the mountains and can be later uplifted by other tectonic movements and erosion. The high pressure and temperature at such depths make the rocks ductile: they can be shaped and moulded without fracturing, which creates folds.

Dynamic metamorphism is typical in the upper crust where pressures and temperatures are lower. Instead of folding, these rocks deform in a brittle manner: when they are under stress due to tectonic movements, they break, fracture and slip. Such behaviour leads to the development of faults and shear zones.

Metasomatism is a special case when rocks are heated by hydrothermal fluids rich in various minerals. As they flow through the rock they alter its composition but don't change the texture.

Shock metamorphism is the result of a meteorite striking the Earth, which affects the bedrock in the proximity of the crater.

Depending on the exact pressure and temperature conditions, some metamorphic rocks get more metamorphosed than others. If you are still able to tell what the parent rock was (its original features are only stretched and distorted, for example), the naming is easy. Was the protolith a gabbro? It's a metagabbro now. Tuff? Metatuff. And so on.

Naming becomes a bit more difficult as we enter high-grade metamorphism that happens at more intense pressures and/or temperatures. The best way to start is with a simple description of the rock's texture. Is the rock stripy (foliated) or does it look more uniform (non-foliated or granular)? Are there any distinct colours? What minerals can you see? Are the minerals oriented in the same direction? This is how we get to names like gneiss and schist. Some igneous protoliths (namely dolerite/diabase and basalt) form a typical range of metamorphic rocks, each characterized by a distinct assemblage of minerals that form under very specific conditions inside the Earth. If you can spot the typical minerals, it's relatively easy to identify the rock.

Some metamorphic rocks include larger mineral grains in their fabric. If the mineral was originally present in the protolith, it is called a porphyroclast (and usually shows signs of being rotated). A porphyroblast, on the other hand, started growing during the metamorphic process.

A lot of rocks contain fluids (mostly water, but also carbon dioxide or sodium chloride) in their mineral structure. During metamorphism, these fluids can become separated from the minerals and often precipitate as veins in fractures.

↓
Mineral veins
cross-cut this pebble

MARBLE

BAKED LIMESTONE

Classification	Granular or foliated
Origin	Contact or regional metamorphism
Grain Size	Fine to coarse
Similar To	Limestone
Name Origin	From Ancient Greek *marmáros* (shining stone, marble)

Marble forms during metamorphism of carbonate sedimentary rocks like limestone and dolomite. With an increase in temperature, the minerals recrystallize and form a mosaic of larger interlocking crystals, giving marble its characteristic marvellous shine. The classic pure white marble forms when there are no impurities in the parent rock – it's just limestone. If there are beds of other rocks (like sandstone, chert or siltstone), they will get metamorphosed alongside the carbonate protolith and create grey and white streaks. Any other minerals present in the limestone will also get recrystallized and give the resulting marble a range of extra colours. Grey is usually caused by graphite, green by olivine, serpentine or tremolite, and blue by diopside.

1

2

1 Marble

2 Forsterite (olivine) marble

LAPIS LAZULI

THE BLUEST ROCK

Classification	Granular or foliated
Origin	Contact metamorphism
Grain Size	Coarse
Similar To	Blueschist, marble
Name Origin	From Latin *lapis* (stone) and Persian *lājward* (blue)

The unmistakable lapis lazuli occurs as lens-shaped bodies in marble. It is thought to form when magma intrudes into limestone and changes its chemistry during contact metamorphism. The deep blue colour is caused by the presence of lazurite, a feldspathoid mineral. The rock is sometimes streaked with white mottled layers or veins of calcite (the dominant mineral in marble and limestone) and golden cubes of pyrite. Thanks to its striking colour, lapis lazuli was used as the source of the precious ultramarine pigment for centuries (though today, this is mostly made synthetically). The lengthy process of grinding, heating and kneading made it more expensive than gold.

HORNFELS

TRANSFORMED BY MAGMA

Classification	Granular
Origin	Contact metamorphism
Grain Size	Fine
Similar To	Gneiss, dolerite
Name Origin	From German *horn* (horn) + *fels* (rock, cliff)

Hornfels forms relatively close to the surface but under quite high temperatures, which can only be achieved around large igneous intrusions. The intense heat of the magma literally bakes the surrounding sedimentary bedrock, forcing the minerals to fuse together. This process turns soft shale and sandstone into the very hard and compact hornfels. Depending on the original rock and its distance from the intrusion, it can be beige to grey and has a typical jagged fracture. A banded hornfels (pictured) forms when alternating beds of shale and sandstone get metamorphosed together.

1 Pelite
2 Psammite

PELITE & PSAMMITE

GENTLE METAMORPHISM

Classification	Foliated
Origin	Regional metamorphism
Grain Size	Fine to medium
Similar To	Slate, phyllite, schist
Name Origin	Pelite from Ancient Greek *pelós* (mud); psammite from Ancient Greek *psámmos* (sand)

In the grand scheme of things, pelite and psammite aren't *that* metamorphosed: they are the deformed equivalents of mudstone and sandstone, respectively. The pressure and temperature they were exposed to were not enough to completely change their appearance and form brand-new minerals. Instead, the original grains were squished together and rotated, often creating distinct layers. Confusingly, the names pelite and psammite used to refer to the unmetamorphosed sedimentary rocks themselves, so some rock formations bear the names metapelite and metapsammite to avoid any doubt about their origin. Various slates and schists are often termed pelitic or psammitic in allusion to their origin.

QUARTZITE

THE HARDEST METAMORPHIC ROCK

Classification	Granular
Origin	Regional or contact metamorphism
Grain Size	Medium to coarse
Similar To	Orthoquartzite
Name Origin	After its main mineral quartz, which comes from Old Church Slavonic *tvrudu* (hard)

Quartzite (also known as metaquartzite) is the metamorphosed version of sandstone rich in quartz (see orthoquartzite, p. 81). As a result of increased pressure and temperature, the quartz in the sedimentary orthoquartzite recrystallizes. The original pore spaces disappear and the mineral grains grow and fuse together to create an interlocking structure. This makes quartzite highly resistant to erosion. A handy way to distinguish between orthoquartzite and quartzite is to look at how it breaks. Orthoquartzite will fracture around the individual grains, while the cracks in quartzite will go right through the merged crystals.

SLATE

METAMORPHOSED SEAFLOOR

Classification	Foliated
Origin	Regional metamorphism
Grain Size	Fine
Similar To	Phyllite, shale
Name Origin	From Middle French *esclate* (splinter)

Slate forms when mudstone, shale or volcanic ash are heated and squeezed. The microscopic quartz, clay and mica grains regrow and orient themselves perpendicular to the direction of the most amount of stress – such alignment is known as slaty cleavage. This layering also means that slate splits very easily along its foliation planes (those are the spaces between mineral layers). This makes it a popular material to use for roof tiles. Its colour is caused by different minerals: grey or black slate contains a lot of graphite, green slate has chlorite and red slate is rich in iron oxides. Pyrite can also be present in black slate as distinct golden or brassy cubic crystals.

PHYLLITE

MORE METAMORPHOSED SEAFLOOR

Classification	Foliated
Origin	Regional metamorphism
Grain Size	Fine
Similar To	Slate, schist
Name Origin	From Ancient Greek *phúllon* (leaf)

Phyllite represents the next step in the metamorphism of mudstone and shale. If slate is metamorphosed just a little bit more, the individual grains of mica begin to grow bigger. This makes them still too small to be seen with the naked eye but large enough to reflect light and give phyllite its characteristic silver-green sheen. Because of the higher pressure and temperature the rock faces during metamorphism, the many layers of mica tend to be irregular with a gentle wavy texture. Much like slate and shale, phyllite also has good fissility, splitting easily along the spaces between the mineral layers.

←
Himalaya Mountains
Asia

The Himalayas began rising around 50 million years ago when the Indian subcontinent collided with the Eurasian tectonic plate. The increased pressure and temperature in the deep parts of the mountain range have led to the formation of a variety of metamorphic rocks, from phyllite and gneiss to amphibolite and granulite.

SCHIST

A ROCK WITH SPARKLE

Classification	Foliated
Origin	Regional metamorphism
Grain Size	Medium to coarse
Similar To	Phyllite
Name Origin	From Greek *schízein* (to split)

As temperature and pressure rise, metamorphosed rocks begin to develop schistosity. This means the individual mineral grains increase in size (up to a point when you can finally see them) and orient themselves parallel to each other to create (often wavy) bands. Schist can contain a whole range of minerals depending on the parent rock's composition and the specific metamorphic conditions. The most common are mica, amphibole, talc, chlorite, actinolite and kyanite. Porphyroblasts (large grains) of garnet, andalusite, staurolite or kyanite can also grow in its metamorphic fabric.

The pale green colour of greenschist (also called greenstone) is due to chlorite, but the rock can also contain actinolite (an amphibole mineral), albite, epidote and quartz. The so-called greenstone belts represent some of the oldest bits of the Earth's oceanic crust – they consist of metamorphosed basalts, komatiites and deep-sea sedimentary rocks.

Blueschist (or glaucophane schist) is a rock formed when oceanic crust is metamorphosed in subduction zones at high pressure and low temperature, and then rapidly uplifted to the surface. The dull blue colour is caused by the amphibole mineral glaucophane – quartz, garnet, pyroxene, epidote or lawsonite are also present in various amounts.

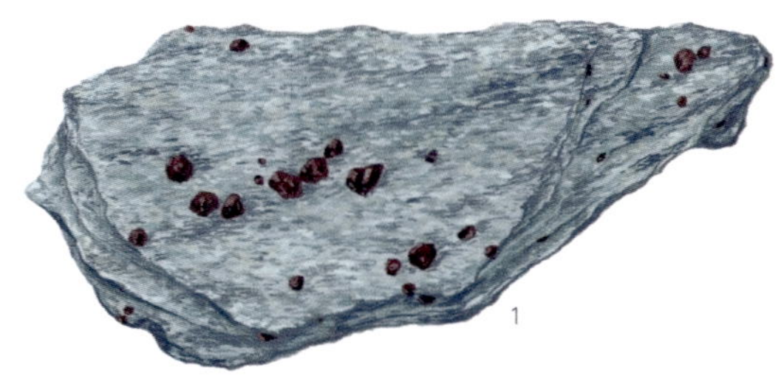

1 Garnet-mica schist
2 Greenschist
3 Blueschist

1 Augen gneiss
2 Orthogneiss

GNEISS

THE OLDEST ROCK ON EARTH

Classification	Foliated
Origin	Regional metamorphism
Grain Size	Coarse
Similar To	Migmatite
Name Origin	From Old High German *gneisto* (spark)

To become gneiss (pronounced 'nice'), a rock is carried relatively deep below the surface where it almost begins to melt. As the pressure and temperature increase, it develops thicker, irregular, alternating bands of dark and pale minerals. This distinct foliation defined by the mineral composition is called gneissosity. The dark layers are typically made of mica, amphibole and/or garnet, while the pale layers contain quartz and feldspars. Depending on the protolith, we can distinguish between orthogneiss (derived from igneous rocks) and paragneiss (derived from sedimentary rocks). Augen gneiss (named after the German word for eyes) contains large lens-shaped grains of feldspar peeking from the darker bands. Some of the Earth's oldest rocks are gneisses, formed during the metamorphism of ancient rocks (like the TTG suite, p. 29).

MIGMATITE

METAMORPHIC OR IGNEOUS?

Classification	Foliated
Origin	Regional metamorphism
Grain Size	Coarse
Similar To	Gneiss
Name Origin	From Ancient Greek *mígma* (mixture)

Migmatite blurs the boundary between what is metamorphic and what is igneous. You can literally see how the parent rock has been melted and folded. The dark parts (made of amphibole and biotite) represent the original rock that was metamorphosed into gneiss. But as the temperature and pressure increase, this protolith begins to melt completely. Migmatite thus shows not only what the most intense metamorphism looks like, but also gives us a glimpse into one of the processes that form new magma. It is called partial melting: when molten, the pale granitic parts flow and form the distinct intermingled and folded texture of migmatite. They also rise through the crust thanks to their lower density and collect in large underground reservoirs, ready to feed volcanic eruptions on the surface.

AMPHIBOLITE

DEEPER METAMORPHISM

Classification	Granular to weakly foliated
Origin	Regional metamorphism
Grain Size	Medium to coarse
Similar To	Diorite, dolerite
Name Origin	After its main mineral amphibole, which comes from Greek *amphíbolos* (ambiguous)

When basalt and deep-sea sedimentary rocks have already been metamorphosed into greenschist (p. 116) but are buried even deeper, they encounter the pressure and temperature conditions needed to form amphibolite. Deep intrusions of dolerite are also often metamorphosed into amphibolite. The characteristic minerals of this rock are (unsurprisingly) amphiboles but plagioclase feldspar, garnet or epidote might also make an appearance. Amphibolite with lots of amphibole is almost completely dark, while more plagioclase feldspar in the mix gives it a salt-and-pepper appearance.

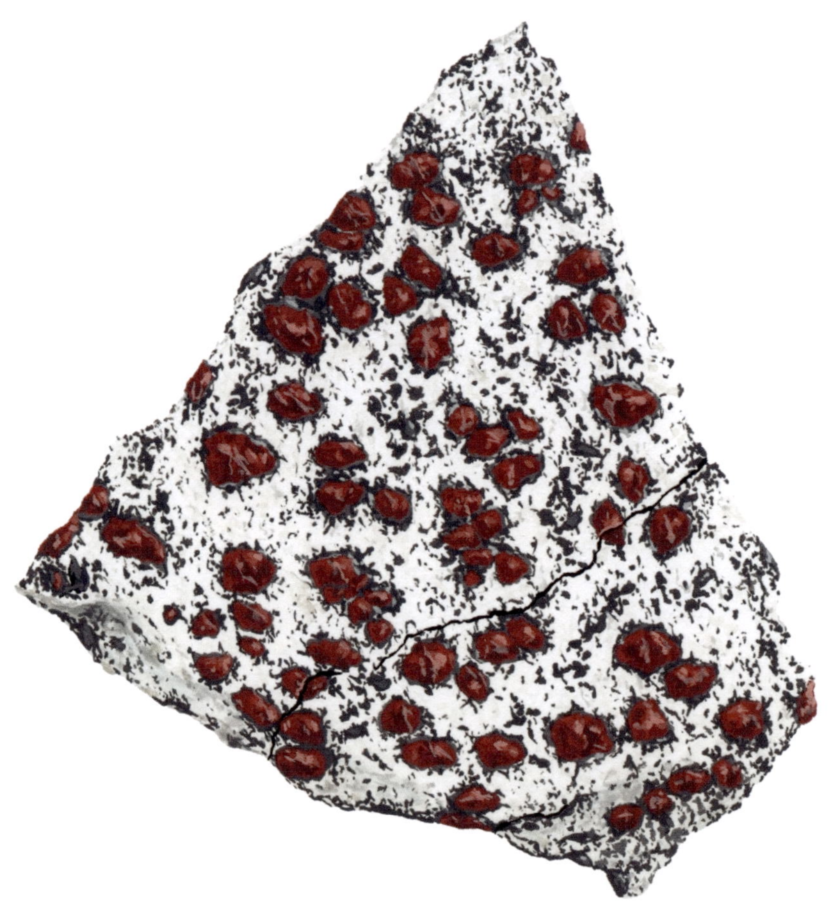

GRANULITE

THE STRONGEST METAMORPHIC ROCK

Classification	**Granular**
Origin	**Regional metamorphism**
Grain Size	**Coarse**
Similar To	**Amphibolite**
Name Origin	**From Latin *grānulum* (a small grain)**

Granulite forms when a rock that has already been metamorphosed into amphibolite finds itself in the deepest parts of mountain ranges. The high temperature is too intense for the water-rich hornblende to exist and it dehydrates. As it loses water, hornblende breaks down into pyroxene minerals. The more resistant plagioclase feldspar, quartz and garnet persist. These minerals not only give granulite a grainy appearance (hence its name) but the lack of water in their structure also makes the rock incredibly strong. Granulite is often found in the oldest parts of continents where it represents the early crust on Earth that was metamorphosed and eventually uplifted.

ECLOGITE

THE DEEPEST-FORMING METAMORPHIC ROCK

Classification	Granular or foliated
Origin	Regional metamorphism
Grain Size	Coarse
Similar To	Garnet peridotite
Name Origin	From Greek *eklogé* (selection)

When basaltic oceanic crust is subducted deep into the mantle, it faces truly extreme pressures. Around 50 kilometres (31 miles) below the surface, it is metamorphosed into eclogite. This rock is composed of only the most resistant minerals that can thrive in deep-Earth conditions: green omphacite (a type of pyroxene), red garnet and blue kyanite. Together, they make eclogite denser than the surrounding mantle, which helps pull the subducting crust downwards. It is very likely that without eclogite, the movement of tectonic plates as we know it would not be possible and the Earth would be a very different place. Via the basalt-to-eclogite transition, water can also be returned into the mantle, which keeps its convection going.

SERPENTINITE

SNAKES IN THE OCEAN

Classification	Granular
Origin	Metasomatism
Grain Size	Variable
Similar To	Epidosite
Name Origin	From Latin *serpens* (creeping thing, serpent)

Serpentinite gets its name from the serpentine group of minerals, ranging from the platy antigorite and lizardite to the fibrous chrysotile (one of the sources of asbestos). These minerals form when seafloor rocks are hydrated. They absorb water and the chemical structure of their original minerals changes. This reaction releases lots of heat and leads to the formation of hydrothermal vents and black smokers on the ocean floor. The best place to see serpentinite on land is in ophiolites. An ophiolite is a portion of the Earth's upper mantle and oceanic crust that was exposed during subduction. A lot of serpentinites are thus peridotites and seafloor basalts that were altered by their exposure to seawater.

EPIDOSITE

GREENING THE SEAFLOOR

Classification	Granular
Origin	Metasomatism
Grain Size	Coarse
Similar To	Unakite
Name Origin	After the mineral epidote, borrowed from French *épidote*

Epidosite is a rock typically found around mid-ocean ridges. As cold seawater percolates downwards through the layers of basalt and dolerite, it gets progressively more heated by the magma below. This heating makes the water move more vigorously and forces it to rise back towards the surface. On the way, it begins to slowly strip away the iron, magnesium and sodium present in the mafic rocks and leaves behind calcium and silicon. Over time, the original minerals are completely replaced by the green epidote and white quartz.

←

Hydrothermal vents
Mid-Atlantic Ridge

Surprisingly, the hot hydrothermal vents and black smokers associated with the creation of serpentinite are also havens for deep-sea life. Colonies of giant tube worms, vent mussels, microbes, shrimp, yeti crabs and other species inhabit these places, benefitting from the heat in the ink-black coldness of the ocean floor. Hydrothermal vents may even be one of the places where life on Earth evolved.

UNAKITE

GRANITE GOES GREEN

Classification	Granular
Origin	Metasomatism
Grain Size	Coarse
Similar To	Epidosite, granite
Name Origin	After the Unaka Range in North Carolina, USA

Hydrothermal fluids can be found circulating through the bedrock in any place with a heat source beneath the surface (like a magma chamber). When these hot, mineral-rich fluids flow through the labyrinthine cracks in granite or gneiss, they begin to replace their sodium-rich plagioclase feldspar with bright green epidote. The salmon-pink alkali feldspar and whitish-grey quartz remain unaffected by this alteration, resulting in the unmistakable multicoloured appearance of unakite.

SKARN

TRANSFORMED LIMESTONE

Classification	Granular
Origin	Metasomatism
Grain Size	Coarse
Similar To	Marble
Name Origin	From Old Norse *skarn* (dung)

When magma intrudes into carbonate rocks and metamorphoses them into marble, the associated fluids begin to influence the bedrock further away. Any carbonates are decarbonized: all their typical minerals are dissolved and replaced with silicates. The same mineral fluids often also leave behind deposits rich in various metal ores (like copper, tin, tungsten and lead) that have been historically mined and extracted. The rest can be made of minerals like talc, wollastonite or even tremolite – a form of asbestos. This part of the rock was useless for the miners, which gave skarn its name.

JADEITITE

A RARE COUSIN OF SERPENTINITE

Classification	Granular or foliated
Origin	Metasomatism
Grain Size	Coarse
Similar To	Serpentinite
Name Origin	After its main mineral jadeite, which comes from Spanish *piedra de ijada* (stone of the side) as it was thought to cure side pains

Jadeitite is a rare rock often associated with blueschist and serpentinite. It is thought to form during subduction in the mantle wedge – a part of the mantle that squeezes its way into the space directly above the down-going plate. The water that escapes from the subducting slab of oceanic crust alters the ultramafic mantle peridotites into serpentinite and, eventually, into jadeitite. Its only mineral component is a clinopyroxene called (surprise) jadeite. The tightly packed and interlocking grains make jadeitite a highly resistant rock, which is often used for tools and decoration.

GREISEN

GRANITE MAKEOVER

Classification	Granular
Origin	Metasomatism
Grain Size	Coarse
Similar To	Granite
Name Origin	From German *greiszen* (to split)

Greisen is a type of granite hydrothermally altered by the very same magma it crystallized from. As the granitic pluton begins to cool down, the last remaining heat drives mineral-rich fluids upwards. They find their way through the cracks in the already solid granite and begin to replace its feldspars with new minerals, most often quartz, mica (muscovite), tourmaline, topaz and fluorite. Some greisens are also rich in metals like tin, tungsten, beryllium and molybdenum. It often occurs as veins that gradually transition into mostly unaltered granite.

MYLONITE

SLIDING ROCKS

Classification	Foliated
Origin	Dynamic metamorphism
Grain Size	Fine to medium
Similar To	Migmatite, gneiss
Name Origin	From Ancient Greek *mylōn* (mill)

Mylonite forms during displacement in the deep parts of large faults. The pressure and temperature around the fault mean that instead of breaking in a brittle manner, the bedrock becomes ductile and starts to flow. To accommodate the immense energy caused by tectonic movement, the minerals shear (strain) and change shape in a process known as dynamic recrystallization. They form new, smaller grains. Some minerals also get rotated and elongated in the direction of the predominant movement, creating mylonite's characteristic streaky appearance.

PSEUDOTACHYLYTE

A FOSSILIZED EARTHQUAKE

Classification	Granular
Origin	Dynamic metamorphism
Grain Size	None
Similar To	Tachylyte, sedimentary breccia
Name Origin	From Greek *pseudés* (false) + *takhús* (swift) + *lutós* (loose, dissolved), pseudo-tachylyte because it resembles the volcanic glass tachylyte

There is only a limited amount of stress rocks can be put under before they fracture and form a fault. During seismic activity, this stress is released when blocks of rock slide at rates of metres per second. Just a small portion of this built-up energy manifests as the ground movement we experience during an earthquake, though. Most of it is released through friction, which heats and melts parts of the bedrock closest to the fault line. This frictional heating creates the characteristic glassy texture of pseudotachylyte (also spelled pseudotachylite).

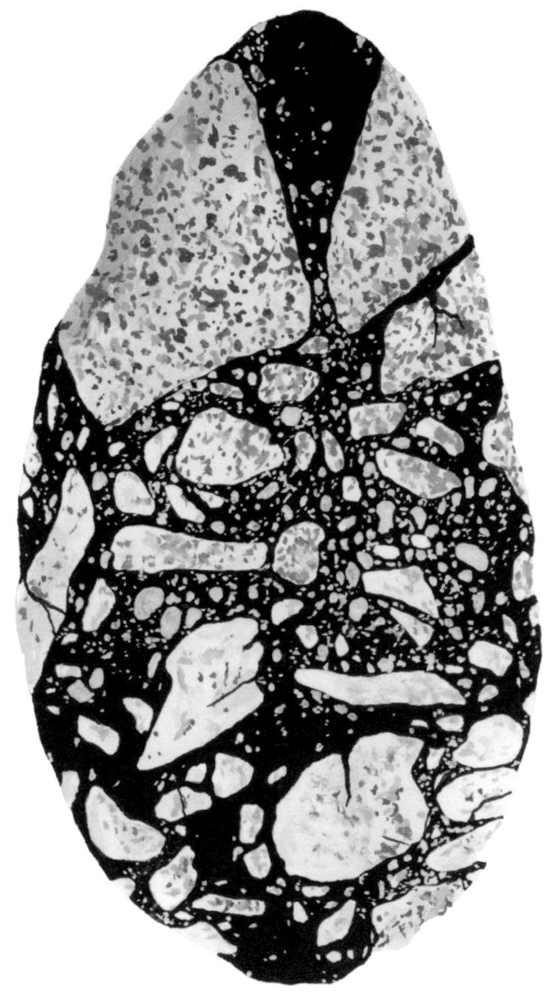

GLACIAL ICE

A DISAPPEARING ROCK

Classification	Granular or foliated
Origin	Glacier
Grain Size	Fine
Similar To	None
Name Origin	Borrowed from French *glace* (ice) and Old English *is* (ice)

Just like any rock, glacial ice is made of minerals. In this case, it's only one mineral: ice. As fresh snow falls on top of a glacier season after season, it builds layers like a sedimentary rock. Over time, the increasing pressure of the weight of new snow metamorphoses the individual snowflakes into denser firn (compacted snow). Add more time and firn loses its bubbles of air and recrystallizes to form transparent glacial ice. It owes its distinct blue colour to the ice crystals themselves: they absorb red wavelengths of the light spectrum and reflect the shorter blue wavelength.

At the same time, glaciers steadily flow and are deformed in the process. Some parts fracture and form deep cracks (crevasses), while others are folded in a similar way to metamorphic rocks. Any impurities in the ice (like tephra and sedimentary grains) can make these folds more noticeable. Today, glacial ice is an endangered rock; the warming seas and atmosphere mean that glaciers are melting at the fastest rate in recorded history.

←

Antarctic ice sheet
Antarctica

The Antarctic continent is covered by the largest ice sheet in the world. The only ice-free places are dry valleys, coastal areas and mountains tall enough to be peaking above the glaciers (these are called nunataks). Continental ice sheets flow towards the Antarctic coasts where they transition into floating ice shelves.

IMPACT BRECCIA

A METEORITE STRIKES

Classification	Granular
Origin	Meteorite impact
Grain Size	Coarse
Similar To	Sedimentary breccia
Name Origin	Borrowed from Italian *breccia* (gravel, rubbish of broken walls)

When a meteorite hits the Earth's surface, it can shatter and melt the bedrock below. The pressure of the impact not only ejects the molten debris into the air but has enough energy to potentially turn carbon into microscopic diamonds. Once the hot glassy material falls down, it mixes with any unmelted sediment and rock fragments. The residual heat then sinters (fuses) everything together to form a solid impact breccia (sometimes also called suevite). Most of the fragments have typical microscopic features (like shock fractures in quartz) that distinguish it from other types of breccia.

TEKTITE

IT'S RAINING GLASS

Classification	Glassy
Origin	Meteorite impact
Grain Size	None
Similar To	Obsidian, glass
Name Origin	From Greek *tēkein* (to melt)

While impact breccia forms at the immediate site of a meteorite strike, tektites can be found strewn far away from it. Smaller debris (called ejecta) thrown into the air by the sheer force of the impact is so light that it keeps flying. As these tiny molten blebs are hurled through the atmosphere, they obtain an aerodynamic shape and are rapidly cooled down to form a glassy substance. Eventually, they rain back onto the ground, forming a symmetrical ejecta blanket around the crater.

Tektites are found in just a few regions on Earth. The dark, circular australites come from southern Australia. They are likely related to similar tektites called indochinites from south-east Asia. The bottle-green moldavites formed around 15 million years ago during a meteorite impact in present-day Germany. They are found around the Vltava (Moldau) river in the south of the Czech Republic.

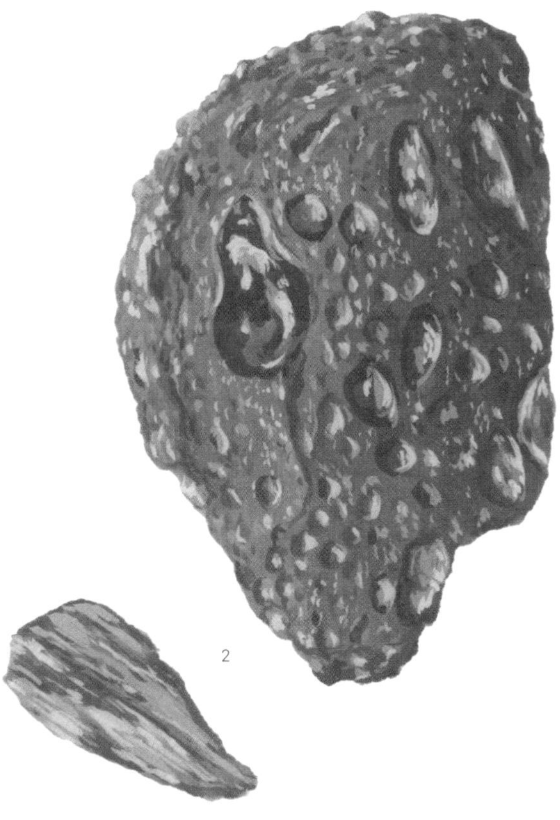

1 Australite

2 Moldavite

ANTHROPIC

ANTHROPIC ROCKS

From Ancient Greek *ánthrōpikós* (related to humans)

Relatively recently (geologically speaking), a new type of process appeared on Earth: human activity. Every day, we are witnessing a major geological event unfold. We move more sediment than all rivers combined. Quarried rocks are often crushed into aggregate and used to build our towns and cities. Minerals and metals mined from rocks make our technology (without them, this book could not have been printed). Petroleum extracted from porous rocks is used in transportation. We've even found ways to create new materials previously unseen on Earth: anthropic rocks. Unlike igneous, volcaniclastic, sedimentary and metamorphic rocks, anthropic rocks exist only because of their value to us. One could argue that they are not proper rocks; they are not formed by natural processes. But when viewed through the lens of deep time, anthropic rocks are not that different from biochemical sedimentary rocks like limestone or coal – they are made by living organisms.

Our lives are literally built on anthropic rocks, but these materials are a double-edged sword. Their production often releases extreme amounts of greenhouse gases and pollutants into the environment. Our way of life has altered the world on a massive scale in a very short amount of time. Acidifying oceans make it harder for marine organisms to build shells; the calcium they need simply dissolves. Our fossil-fuel emissions warm the atmosphere, melt ice sheets and raise sea levels. The rate of extinction of many non-human species is well above the normal background rates. The climate and biodiversity crises affect humans too – especially the most vulnerable communities.

Suddenly, geology merges with anthropology, sociology, biology, ecology, archaeology, economy, philosophy, politics, engineering... Anthropic rocks remind us that humans are part of a global system and everything we do – or don't do – has an effect.

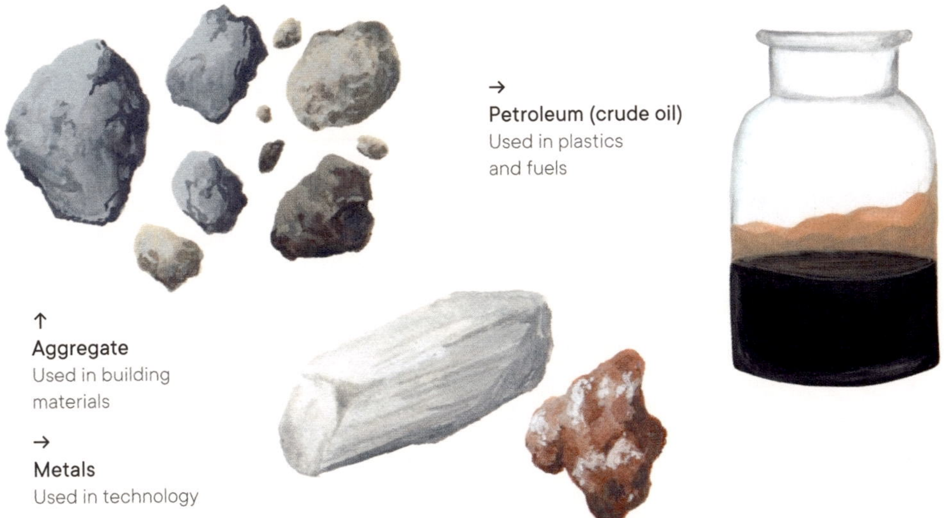

A NEW ERA FOR ROCKS

Human traces are likely to be visible in the geological record for millions of years to come. Most surface features will be broken apart, buried, recycled and transformed by erosional and tectonic processes. Underground tunnels, shafts and boreholes have a much higher chance of being preserved. There will be our own fossils, of course, as well as the incredible amount of bones of domesticated animals. The future fossil record of today will probably be dominated by chickens.

At the same time, rocks will keep forming without any human input. Just as they always have. Rivers will escape their straight embankments, dams will breach, seas will advance and retreat. Geology will keep happening. However, human-made materials are slowly but surely finding their way into rocks forming right now. At the end of this section, I engage in partially speculative geology to think about such human-influenced rocks. Some of them have already been found. Others will take millions of years to form. We don't know what will happen to anthropic rocks in the depths of geological time because most of them have never been present on Earth before. How will concrete change when it gets metamorphosed? What will microplastics do to deep sea sediments? How will aluminium cans cope with being entombed in rock for millions of years?

POTTERY

THE OLDEST ANTHROPIC ROCK

Classification	Ceramic
Environment	Urban areas
Grain Size	Fine
Similar To	Claystone, fired brick
Name Origin	From Middle French *poterie* (the art of making pots)

Pottery encompasses all ceramics not used for building purposes. The earliest known pottery – a figurine called the Venus of Dolní Věstonice – was sculpted around 29,000 years ago in the present-day Czech Republic. All pottery is made by shaping and firing a mixture of clay (usually kaolin), mica, quartz, feldspar and sometimes chamotte (crushed pottery fragments). The heat of the kiln drives off water present in the minerals and forces the clay particles to fuse. Reddish earthenware (also called terracotta) is fired at the lowest temperatures (below 1,200°C or 2,190°F) and easily absorbs water. The higher firing temperatures of stoneware (up to 1,300°C or 2,370°F) and porcelain (up to 1,400°C or 2,550°F) turn the clays into glass, making them more resistant and watertight. Pottery can be decorated with a layer of colourful glaze, a protective, water-repellent glassy coating made from a variety of materials.

1

2

3

1 Glazed stoneware
2 Earthenware
3 Porcelain

FIRED BRICK

HUMAN-MADE METAMORPHISM

Classification	Ceramic
Environment	Urban areas
Grain Size	Fine
Similar To	Claystone, pottery
Name Origin	From Middle Dutch *brycke* (brick)

Geologically speaking, a fired brick is a human-made, slightly metamorphosed sandy claystone. It starts its life as a mixture of quartz sand, clay, limestone and iron oxides that is shaped and fired in a kiln to produce a durable material. The colour depends on its composition (ochre bricks contain more lime, while redder bricks have more iron oxide) and the firing temperature (lower temperatures for dark red, higher temperatures for purple to brown). Bricks are normally fixed in walls using lime mortar (made from limestone), which can sometimes continue holding them together even after the original structure has collapsed.

GLASS

HUMAN-MADE OBSIDIAN

Classification	Composite material
Environment	Urban areas
Grain Size	None
Similar To	Obsidian
Name Origin	From Old English *glæs* (glass)

Glass can have variable compositions based on its desired properties but is usually made using quartz, sodium carbonate (soda ash), limestone and recycled glass. This mixture is heated to nearly 1,600°C (2,910°F) – higher than any lava erupting today – which melts and fuses the ingredients together. The pliable glass can then be moulded or blown into any shape before being cooled down relatively slowly to prevent it from cracking. This cooling process creates an amorphous (non-crystalline) solid similar in composition and texture to obsidian (p. 50). Just like its volcanic equivalent, glass also devitrifies over time.

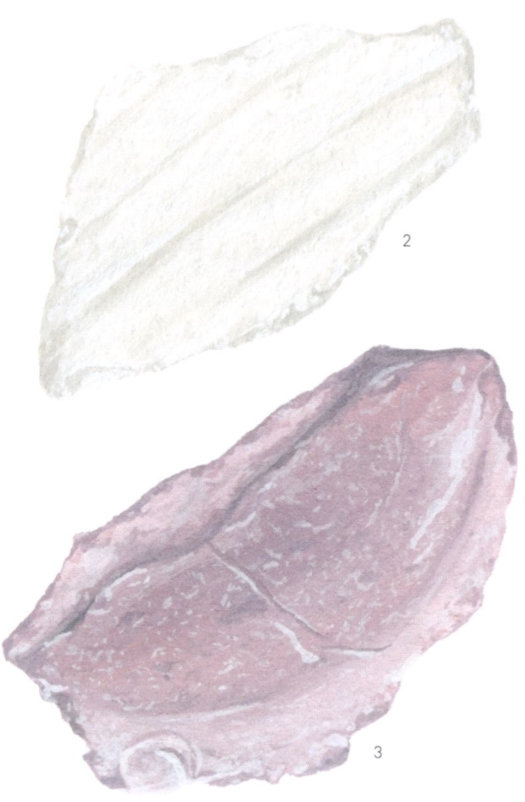

1 Chromium-coloured glass
2 Clear glass
3 Manganese-coloured glass
4 Cobalt-coloured glass

ALLOY

THE MEETING OF METALS

Classification	Metal mixture
Environment	Mostly urban areas
Grain Size	Fine
Similar To	Industrial slag, meteorite
Name Origin	From French *aloier* (to combine metals)

Alloy is a general term for a mixture of a metal with another chemical element. The first human-made alloy was bronze (copper + tin), which pushed human civilization in a whole new direction; the first urban societies date to the Bronze Age (around 3500 BCE). Since then, we have found ways to create more alloys, such as steel (iron + carbon), stainless steel (iron + carbon + chromium), brass (copper + zinc), pewter (tin + antimony + copper), rose gold (gold + copper + silver) and white gold (gold + silver). The metals needed for alloys are mined from ores. These are rocks rich in metals which form in a number of ways, including from magma, through hydrothermal circulation and sedimentary processes. Not all alloys are anthropic, though: prime examples are the iron-nickel alloy found in the Earth's core (and some meteorites) and electrum (gold + silver).

1

2

3

1 White gold
2 Bronze
2 Steel

INDUSTRIAL SLAG

A METEORITE LOOK-ALIKE

Classification	Byproduct
Environment	Urban areas but can be found anywhere
Grain Size	Fine
Similar To	Alloy, meteorite, vesicular basalt
Name Origin	From Middle Low German *slagge* (to forge metal but also rainy weather)

If you think you found a meteorite, the chances are quite high that you did not (sorry). It is more likely to be industrial slag – a byproduct of steel production. During the manufacturing process, raw iron ore is mixed with coal and limestone to make the mixture more fluid and remove any impurities that could affect the final product. Like a lot of meteorites, slag is magnetic, due to its high iron content. Unlike a meteorite, though, slag has irregular surfaces and vesicles (gas bubbles) of many shapes and sizes (sometimes infilled with shiny iron). On the other hand, the surface of a meteorite is normally smoothed out by flying through the atmosphere.

CONCRETE

THE MOST WIDESPREAD ANTHROPIC ROCK

Classification	Composite material
Environment	Urban areas
Grain Size	Fine to coarse
Similar To	Limestone
Name Origin	From Latin *concrēscere* (to grow together)

Concrete is the most abundant anthropic rock – it has been in use for millennia. Its production is quite simple: cement binder (a mixture of lime, clay and gypsum) is combined with aggregate (usually sand, gravel, crushed rock or even recycled concrete) and water. This creates a liquid slurry that, once mixed, starts to crystallize and harden. The chemical reaction releases carbon dioxide, which sometimes leaves behind vesicle-like holes. Many concrete structures are reinforced with steel bars inside, providing additional support. Roman concrete incorporated volcanic ash called pozzolana, which allowed any cracks to heal by themselves.

ASPHALT CONCRETE

ROADS FULL OF PLANKTON

Classification	Composite material
Environment	Urban areas
Grain Size	Fine to medium
Similar To	Coal
Name Origin	From Greek *ásphaltos* (asphalt, bitumen) and Latin *concrēscere* (to grow together)

Covering road surfaces all around the world is asphalt concrete (also known as tarmac), prepared and laid down wherever needed. It is a mixture of aggregate (sand, gravel and crushed rocks) and bitumen (also called asphalt or pitch), a sticky substance obtained by distilling petroleum (aka crude oil). Petroleum – also used to make plastic – is a fossil fuel: a natural material formed when dead plankton, algae and plants get buried deep in oxygen-poor ocean or lake sediments. Instead of decaying, they are slowly transformed by heat and pressure into kerogen, oil and, ultimately, gas.

SCAGLIOLA

HUMAN-MADE MARBLE

Classification	Composite material
Environment	Urban areas
Grain Size	Variable
Similar To	Marble
Name Origin	From Italian *scagliuola* (a small chip of gypsum)

When plaster of Paris (gypsum mixed with water), glue and pigments are combined together, they form a thick paste. Scagliola became popular during the Italian Baroque period in the 17th century as a cheap alternative to decorative marble (p. 108). Colourful scagliola mixes were combined and layered to create intricate swirling patterns reminiscent of the layers in marble – often in bright colours you wouldn't see in the real metamorphic rock.

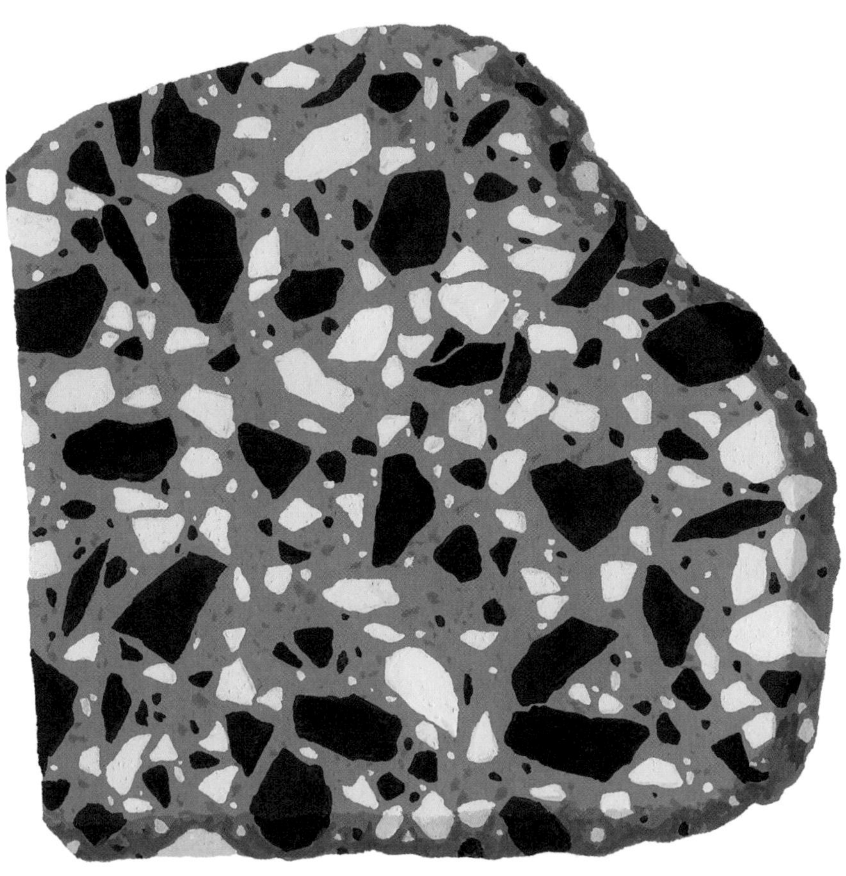

TERRAZZO

HUMAN-MADE BRECCIA

Classification	**Composite material**
Environment	**Urban areas**
Grain Size	**Variable**
Similar To	**Sedimentary breccia**
Name Origin	**From Italian *terrazzo* (terrace, balcony)**

A mosaic-like surfacing material, terrazzo has been traditionally made from a mixture of cement and an aggregate (usually marble left over from quarrying large blocks). This classic version is not too different from a sedimentary breccia (p. 83) held together by a carbonate matrix. However, new forms have emerged. The lime-based cement can be substituted with epoxy resin (which can be of any colour depending on additives) and the aggregate can be of any composition. The mixture is poured or cast and once it hardens, terrazzo is normally polished to create a smooth surface.

PLASTIC

ANCIENT PLANKTON
EVERYWHERE

Classification	Synthetic material
Environment	Everywhere
Grain Size	None
Similar To	Petroleum
Name Origin	From Ancient Greek *plastikós* (able to be moulded)

Plastics are mostly made from crude oil, a substance formed by the quick burial and decomposition of dead plankton, algae and plants in environments low in oxygen (such as deep oceans and lakes). The production of plastic materials has been rapidly increasing since the 1950s. Polyethylene terephthalate (PET or polyester), polyethylene (PE), polypropylene (PP), polystyrene (PS) and polyvinyl chloride (PVC) are the most common types, each with its specific uses. Plastics are so widespread simply because they make our lives much easier. They can be made into virtually any shape possible, don't corrode like most metals and resist microbes. This also means that once they enter the environment, plastic products don't decompose but break down into smaller pieces known as microplastics (p. 158).

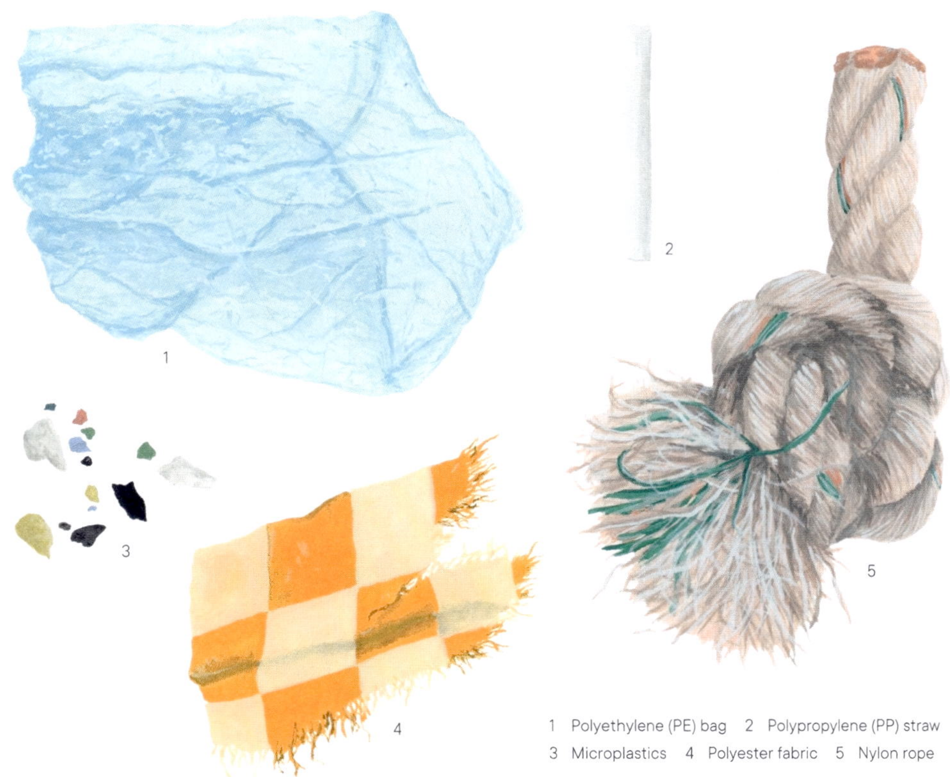

1 Polyethylene (PE) bag 2 Polypropylene (PP) straw
3 Microplastics 4 Polyester fabric 5 Nylon rope

TRINITITE

A WITNESS TO VIOLENCE

Classification	**Byproduct**
Environment	**Nuclear test sites**
Grain Size	**None**
Similar To	**Tektite**
Name Origin	**After Trinity, the code name given to the first nuclear bomb test**

Some anthropic rocks were created by accident and trinitite is a prime example. Its formation can be tracked down to an exact minute – 5:29 AM, 16 July 1945. At that time, the first nuclear detonation took place near Alamogordo in New Mexico, USA. The effects of the intense heat of the blast were comparable to a meteorite strike. It ejected the desert sand (made mostly of quartz and feldspars) into the fireball, rapidly heated it, melted the individual grains and fused them together to form a greenish glass. It is likely that all nuclear bomb detonations formed a similar material.

CALTHEMITE

STALACTITES IN CITIES

Classification	Secondary deposit
Environment	Urban areas
Grain Size	Fine
Similar To	Travertine
Name Origin	From Latin *calx* (lime) + Ancient Greek *thémai* (to place or lay down)

Calthemites are the stalagmite-like straws and crusty flow layers growing from tunnel ceilings, under bridges, on walls, pipes and really anywhere where there is an abundance of concrete. When water flows through microscopic cracks in concrete, it partially dissolves the lime it contains. This enriches the water in calcium hydroxide. Once this liquid comes in contact with air, it reacts with carbon dioxide and precipitates pure calcium. Calthemites are most often white but can also have a rusty colour if the concrete is reinforced with iron, or turquoise if the water encounters copper. The thin straws can grow up to 2 millimetres per day (much quicker than travertine stalactites, p. 93) due to the higher alkalinity of the water.

PLASTIGLOMERATE

ROCKS IN PLASTIC

Classification	Trace fossil
Environment	Coast
Grain Size	Various
Similar To	Conglomerate, beachrock
Name Origin	From Ancient Greek *plastikós* (able to be moulded) + Latin *glomerāre* (to roll into a ball)

Any plastic waste will eventually find its way into the environment. The majority of our plastic debris ends up in the oceans if it wasn't discarded there in the first place. The lightweight, free-floating particles are not only easily mistaken by marine organisms for food, but also accumulate around the coasts, gradually embedding themselves in the sediment. Plastiglomerate forms when such polluted sediment is heated by fires, bonfires or lava. The plastic melts and creates a kind of matrix that glues together the surrounding sediment grains, rock fragments, fossils and even other bits of plastic debris.

PYROPLASTIC

FLOATING PLASTIC PEBBLES

Classification: Trace fossil
Origin: Coast
Grain Size: None
Similar To: Various beach pebbles
Name Origin: From Ancient Greek *pûr* (fire) + *plastikós* (able to be moulded)

Similarly to plastiglomerate, pyroplastic is formed by fire. When plastic debris is heated, it melts and takes the form of roundish, pebble-like objects. It often has various bubbles and fractures, and is easily mistaken for any rock smoothed by the waves. However, unlike non-anthropic rocks (apart from pumice, p. 49), pyroplastic floats on water thanks to its low density. It can re-enter the ocean and continue breaking down into microscopic particles. It is usually whitish to blackish but can be of any colour depending on the composition of the original plastic material.

←
Great Pacific Garbage Patch
Pacific Ocean

Trapped in a circling gyre of ocean currents are trillions of pieces of plastic debris. They range from ghost fishing nets (lost or intentionally discarded at sea) to waste coming from the countries around the Pacific coasts, and microplastics created as large pieces break down. They often end up washed up on Kamilo Beach on The Big Island of Hawaii, thought to be one of the most plastic-polluted coastal areas in the world. This is where the first plastiglomerates were found and described in 2014.

TECHNOFOSSIL

THE TRACES OF OUR LIVES

Classification	Trace fossil
Environment	Various
Grain Size	Various
Similar To	Various
Name Origin	From Ancient Greek *tékhne* (craft, skill) + classical Latin *fossilis* (obtained by digging)

Technofossils are the footprints of our existence. Plastics in coastal areas get embedded in the carbonate sediment: over geological timescales, the hydrocarbons within them will leach, leaving behind an imprint or a cast of the original object. But technofossils aren't restricted to coasts. Even today we are finding Stone Age tools buried in technosols (soils modified by humans). Coins intentionally pressed into molten lava were popular as souvenirs in Naples in the 17th to 19th centuries. Explosive volcanic eruptions cover whole towns and encase them as ash turns to rock. In landfills our waste is protected from erosion by layers of sediment (and more waste). Radioactive waste goes even deeper: some of it is stored in underground geological repositories where it can sit undisturbed for hundreds of thousands of years. These are just a few examples. Anything used by humans that leaves a mark in rocks can become our trace fossil.

1

2

3

4

1 Coin in lava 2 Pen cap cast in limestone
3 Fossilized landfill deposit 4 Mosaic in tuff

ANTHROPOQUINA

ANTHROPIC ROCKS AS GRAINS

Classification	Trace fossil, bioclastic sediment
Origin	Coast, continent
Grain Size	Various
Similar To	Sandstone, limestone, beachrock
Name Origin	From Ancient Greek *ánthrōpikós* (related to humans) + *coquina* (a type of shelly limestone)

Anthropoquina is a sedimentary rock with various human-made materials (including plastic, bricks, glass, alloys and concrete) embedded within it. Unlike a plastiglomerate (p. 152), it doesn't need any heat to form because the debris is incorporated through natural sedimentation processes just like any other sedimentary particle. Plastic becomes clastic. Grain by grain, human-made objects are buried. Once the material is deep enough under the surface, it can be cemented together and form a solid anthropoquina. The chances of it being preserved in the geological record rapidly increase. Some quickly forming beachrocks (p. 86) have been found to contain anthropic materials.

MICROPLASTIC SHALE

PLASTIC RETURNS TO ITS SOURCE

Classification	Trace fossil, bioclastic sediment
Environment	Deep ocean
Grain Size	Fine
Similar To	Mudstone, shale
Name Origin	From Ancient Greek *mikrós* (small) + *plastikós* (able to be moulded)

Plastic debris that bobs up and down in the oceans is degraded by ultraviolet radiation from the sun. Large pieces of plastic become brittle and begin to break down into microplastics (anything smaller than 5 millimetres). Other plastics enter the ocean already small, like microfibres from clothing, tyre particles and flakes of synthetic paint. All these minuscule plastics eventually settle in the muddy sediment at the bottom of the ocean. If you managed to pick up a handful of ocean mud today and look at it under a microscope, it would be full of plastic pieces. Over millions of years, the muddy sediment gets buried and turns into a 'plastishale'. The hydrocarbons in plastic will likely get released by the increased heat and perhaps even form new oil deposits, going full circle to where they started.

URBANOGLOMERATE

THE FUTURE ROCK OF OUR CITIES

Classification	Trace fossil, bioclastic sediment
Environment	Shallow to deep ocean
Grain Size	Various
Similar To	Conglomerate
Name Origin	From classical Latin *urbānus* (belonging to a city) + *glomerāre* (to form into a ball)

As sea levels keep rising, the most vulnerable settlements nearest to seas will eventually be drowned. The concrete foundations, steel supports, asphalt roads, glass windows, kilometres of electric cables, tonnes of plastic and anything we leave behind will be slowly broken down and covered by marine sediment and protected from extensive erosion. Over millions of years, as this urban layer is covered by more sediment, it will be consolidated, cemented and turned into rock. A conglomerate of sorts, but with a distinct abundance of anthropic materials. Glass will devitrify and turn into an opaque material not unlike snowflake obsidian (p. 50). Bricks might become clays again. Metals will corrode in the presence of oxygen; in oxygen-poor waters, iron might turn into pyrite. Even though some of the materials will be changed beyond recognition, their essence will persist.

A FINAL WORD

Next time you pick up a rock, remember that it is not just a rock. You are holding a piece of the Earth's history – your history. And you might be the first human ever to touch it. With this in mind, take a moment to slow down. Try experiencing the world through this rock. What actions, states and events does it contain? Let it take you on a journey to the deep past. Imagine where it was an hour ago, before you met it (probably in the same place, to be honest). Now, feel deeper. Where was it yesterday? Where was it ten, a hundred, a thousand, a million, a hundred million years ago? When was the last time it felt the warmth of the sun? How did it move through the Earth system? Think of the many changes it has gone through over time. What was it before it formed into the rock that is in your hand?

Then, return to the present day. Look at the landscape around you. Notice the plants, fungi and animals; many of them live in this exact spot because of the specific rock type. Even the local climate is a result of geological processes. The piece of crust you are standing on has been moved around by tectonic forces only to (temporarily) end up right here, right now. You are witnessing a snapshot in the evolution of the Earth. Past landscapes, with their own ecosystems, are right beneath your feet. Your rock has likely been a part of them. It is going to be in landscapes that are yet to form.

Geologists like to say that every rock tells a story, but it is perhaps more accurate to say that every rock *is* a story. When we pay attention to a rock – think *with* and *through* it, not just *about* it – our stories converge for a brief moment in the vastness of deep time.

→
Holding deep time
I found this rock at 11:01 on 13 December 2024, at Carlingheugh Bay, Arbroath, Scotland. This porphyry was molten magma with growing crystals, then erupted as lava, buried by sediment, eroded by rivers and glaciers, and tumbled by ocean waves.

GEOLOGICAL GLOSSARY

Terms in *italics* featured elsewhere in glossary

Aggregate
Crushed *rocks* used in the production of a range of materials

Amygdale
A *vesicle* filled with secondary hydrothermal *minerals* (quartz, calcite, zeolites); also spelled amygdule

Anthropic
Rocks made by humans or influenced by human activity

Batholith
A complex underground body of *magma* covering over 100 square kilometres (38.6 square miles)

Bed
A layer of *sedimentary* or *volcaniclastic rock* with defined upper and lower boundaries thicker than 1 centimetre (0.4 inches); also called a stratum (plural strata)

Bedding plane
The upper and lower surface of a *sedimentary bed*

Bedrock
The solid foundation of *rock* that lies beneath loose surface *sediment* and *soil*

Brittle deformation
A process that leads to fractures and *faults* in *bedrock*

Cement
Minerals that bind the *grains* of *sediments* together to form a *sedimentary rock*; also a human-made material made from lime and clay

Clast
Any *rock* or *mineral grain* that makes a clastic *sedimentary* rock

Cleavage
A plane of weakness between *mineral grains* in *metamorphic rocks*

Cross-bedding
Layers in a *sedimentary rock* at an angle to the main *bedding plane* which indicate the direction of movement

Crystal
A solid material with a defined chemical composition and shape

Deep time
The billions of years of the Earth's geological history; it includes all the events that have ever happened and are still happening

Devitrification
The process of changing from glass into a crystalline structure

Diagenesis
The chemical, physical and biological processes *sediments* undergo as they turn into solid *rock*

Ductile deformation
A process that leads to *folding* of *bedrock*

Dyke
A vertical body of *igneous rock* that *intruded* into older *bedrock*; also spelled dike

Earthquake
Ground movement caused by the release of energy in the crust during *tectonic* processes

Effusive eruption
A volcanic eruption of *magma* with a low amount of water and gases; it normally produces *lava* flows

Erosion
The removal and transport of *rocks*, *sediments* and *soil* by water, ice, wind and gravity

Explosive eruption
A volcanic eruption of *magma* which contains a lot of water and gases that build up pressure; it normally produces ash

Fault
A fracture that leads to displacement in *bedrock*

Felsic
An *igneous rock* which contains *minerals* rich in silica (quartz, feldspar)

Fissility
The ability of *rocks* to split into thin sheets

Fold
A horizontal *rock* layer which was bent by pressure and temperature

Foliation
Repetitive layers in *metamorphic rocks* such as slate, schist and gneiss

Fossil
The remains of a plant, animal or fungus whose tissues were replaced by *minerals*

Fractional crystallization
A process when *minerals* crystallize from *magma* in stages

Geode
A roughly spherical cavity in a *rock* filled with *mineral crystals precipitated* from *hydrothermal fluids*

Grain
A *mineral crystal* (in *igneous* and *metamorphic rocks*) or a fragment of a rock or a mineral (in *sedimentary* rocks)

Groundmass
The fine-grained material in an *igneous* or *sedimentary rock* that may surround larger *crystals* or *grains*; also called *matrix*

Hydrothermal fluids
Hot, *mineral*-rich groundwater that circulates in the Earth's crust

Igneous
Rocks formed when molten *magma* or *lava* cool down and crystallize

Intermediate
An *igneous rock* with a composition between *mafic* and *felsic*

Intrusion
A body of *igneous rock* that cooled down and crystallized below the surface (see also *dyke* and *sill*)

Lamination
A layer of *sedimentary* or *volcaniclastic rock* with defined upper and lower boundaries thinner than 1 centimetre (0.4 inches)

Lava
A molten mixture of solid *crystals* and liquid *rock* erupted from a *volcano*

Lithic
Made of, or relating to, *rock* (from the Ancient Greek *líthos* meaning 'stone')

Lithification
The process of turning loose *sediment* into solid *rock*

Lithosphere
The crust and the uppermost part of the mantle; it is split into *tectonic* plates

Mafic
An *igneous rock* which contains *minerals* rich in iron and magnesium (olivine, pyroxene)

Magma
A molten mixture of solid *crystals* and liquid *rock* found beneath the Earth's surface

Matrix
The fine-grained material in an *igneous* or *sedimentary rock* that may surround larger *crystals* or *grains*; also called *groundmass*

Metamorphic
Rocks changed by pressure and/or temperature

Metamorphism
The processes that change a *rock's* texture and *mineral* composition as a result of increased temperature and/or pressure

Microplastic
A fragment of plastic less than 5 millimetres in size

Mineral
A naturally occurring solid with a defined chemical composition and *crystal* structure

Monomict
A *clastic sedimentary rock* whose *grains* have the same composition

Ophiolite
Fragment of ancient ocean floor pushed onto land by *tectonic* movements

Ore
A *rock* with a high quantity of metals and/or other economically valuable *minerals*

Orogeny
The process of mountain building during the collision of *tectonic* plates

Outcrop
Bedrock visible at the Earth's surface

Partial melting
The process of melting parts of pre-existing *rocks* to create *magma*

Phenocryst
A distinct larger *mineral* in the *groundmass* of an *igneous rock*

Pillow lava
The rounded, pillow-like shapes formed when *lava* erupts underwater

Plankton
A collection of small organisms that drift in water currents; they form the base of ocean and freshwater food webs

Pluton
A large *igneous intrusion*

Polymict
A *clastic sedimentary rock* whose *grains* have different compositions

Porosity
The presence of empty spaces within a *rock*

Porphyroblast
A large *mineral* formed during *metamorphism*

Porphyroclast
A large remnant of the *protolith* in a *metamorphic rock*

Precipitation
A process when *minerals* dissolved in a fluid come out of solution and form solid *crystals*

Protolith
The original *rock* before it was *metamorphosed*; also called parent rock

Pyroclast
A fragment of *lava* ejected into the air during an *explosive eruption*

Regolith
A loose mixture of *rock* fragments and dust that covers solid *bedrock*

Rock
A solid material made of one or more *minerals*

Sediment
Loose material (*mineral* and *rock grains*, remains of plants and animals) transported by *erosion*

Sedimentary
Rocks formed when *grains* of *sediment* are *cemented* together, through chemical *precipitation* or by living organisms

Siliceous ooze
A type of *sediment* made of the remains of *plankton* such as diatoms and radiolarians

Sill
A horizontal body of *igneous rock* that *intruded* into older *bedrock*

Soil
A mixture of *mineral* and *rock* fragments, organic matter, living organisms, water and gases; also called earth

Stone
A term for a *rock* most often used when it is a building or creative material

Subduction
A boundary between two *tectonic* plates where one slides under the other and is recycled in the mantle; this process often creates *volcanoes* and mountains

Tectonic
Related to the structure, movements and processes of the Earth's crust; from the Greek *tektonikós* (pertaining to building)

Tephra
Fragments of *lava* (ash, lapilli, bombs) which blanket the landscape after an *explosive eruption*

Trace fossil
Preserved remains of biological activity, such as burrows, footprints or feeding traces; also called ichnofossil

Type locality
The area where a specific *rock, mineral, fossil* or geological formation was scientifically described for the first time

Ultramafic
An *igneous rock* which contains even more iron and magnesium than a *mafic* igneous rock

Vein
A fracture in a *rock* infilled with secondary hydrothermal *minerals* (quartz, calcite, zeolites)

Vesicle
A hole in a volcanic *igneous rock* left by *volatile* gases and liquids escaping from the *magma* during an eruption

Viscosity
The measure of how easily a fluid flows. Fluids with low viscosity move easily while highly viscous fluids are thicker and more resistant to flow

Volatiles
Gases and liquids (most often carbon dioxide and water) dissolved in *magma* at depth

Volcaniclastic
Rocks formed when *pyroclasts* erupted from *volcanoes* are *cemented* together

Volcano
An opening or vent in the Earth's surface where *magma* rises through a network of cracks in the *bedrock* and erupts as *lava* or ash

Weathering
The chemical, physical and biological processes that lead to the breakdown of *rocks*

Xenocryst
A piece of *rock* of a different composition inside an *igneous* rock

Xenolith
A *mineral* in an *igneous rock* that did not crystallize from the same *magma*

FURTHER READING & RESOURCES

INTRODUCTION TO GEOLOGY

Bjornerud, M. *Geopedia: A Brief Compendium of Geologic Curiosities*, Princeton University Press, Princeton NJ, 2022.

Bonewitz, R. L. *Rocks & Minerals: The Definitive Visual Guide*, DK, London, 2023.

Gordon, H. *Notes from Deep Time: A Journey Through Our Past and Future Worlds*, Profile Books, London, 2021.

Pellant, C. and Pellant, H. *Rocks and Minerals*, DK, London, 2021.

Rogers, S. L., Lau, L., Dowey, N., Sheikh, H. and Williams, R. 'Geology uprooted! Decolonising the curriculum for geologists', *Geoscience Communication*, 5 (2022), pp. 189–204.

Zalasiewicz, J. *The Planet in a Pebble: A journey into Earth's deep history*, Oxford University Press, Oxford, 2020.

IGNEOUS ROCKS AND VOLCANOES

Andrews, R. G. *Super Volcanoes: What They Reveal about Earth and the Worlds Beyond*, W. W. Norton & Co, New York NY, 2021.

Gill, R. and Fitton, G. *Igneous Rocks and Processes: A Practical Guide*, 2nd edn, Wiley-Blackwell, Hoboken NJ, 2022.

Sigurdsson, H. (ed). *The Encyclopedia of Volcanoes*, 2nd edn, Academic Press, Cambridge MA, 2015.

SEDIMENTARY ROCKS AND FOSSILS

Boggs Jr., S. *Principles of Sedimentology and Stratigraphy*, 5th edn, Pearson, London, 2016.

Halliday, T. *Otherlands: A World in the Making*, Penguin Books, London, 2022.

Leeder, M. R. *Sedimentology and Sedimentary Basins: From Turbulence to Tectonics*, 2nd edn, Wiley, New York NY, 2011.

Panciroli, E. *The Earth: A Biography of Life: The Story of Life On Our Planet through 47 Incredible Organisms*, Greenfinch, London, 2022.

METAMORPHIC ROCKS

Bucher, K. *Petrogenesis of Metamorphic Rocks*, 9th edn, Springer Nature, Switzerland, 2023.

Frost, B. R. and Frost, C. D. *Essentials of Igneous and Metamorphic Petrology*, 2nd edn, Cambridge University Press, Cambridge, 2019.

ANTHROPIC AND FUTURE ROCKS

Corcoran, P. L., Moore, C. J. and Jazvac, K. 'An anthropogenic marker horizon in the future rock record', *GSA Today*, 6 (2014), pp. 4–8.

De-la-Torre, G. E., Dioses-Salinas, D. C., Pizarro-Ortega, C., Pizarro-Ortega, I. and Santillán, L. 'New plastic formations in the Anthropocene', *Science of the Total Environment*, 754 (2021).

Eby, N., Hermes, R., Charnley, N. and Smoliga, J. A. 'Trinitite—the atomic rock', *Geology Today*, 26 (5) (2010), pp. 180–5.

Farrier, D. *Footprints: In Search of Future Fossils*, Fourth Estate, London, 2020.

Fernandino, G., Elliff, C. I., Francischini, H. and Dentzien-Dias, P. 'Anthropoquinas: First description of plastics and other man-made materials in recently formed coastal sedimentary rocks in the southern hemisphere', *Marine Pollution Bulletin*, 154 (2020).

Harkness, R. (ed). *An Unfinished Compendium of Materials*, The University of Aberdeen, Aberdeen, 2017.

Turner, A., Wallerstein, C., Arnold, R and Webb, D. 'Marine pollution from pyroplastics', *Science of The Total Environment*, 694 (2019).

Waters, C. N., Graham, C., Tapete, D., Price, S. J., Field, L., Hughes, A. G. and Zalasiewicz, J. 'Recognizing anthropogenic modification of the subsurface in the geological record', *Quarterly Journal of Engineering Geology and Hydrogeology*, 52 (2019), pp. 83–98.

Zalasiewicz, J., Williams, M., Waters, C. N., Barnosky, A. D. and Haff, P. 'The technofossil record of humans', *The Anthropocene Review*, 1 (1) (2014), pp. 34–43.

RECONNECTING WITH ROCKS

Allen, R. *Weathering: How the Earth's deep wisdom can help endure life's storms*, Ebury Press, London, 2024.

Bjornerud, M. *Timefulness: How Thinking Like a Geologist Can Help Save the World*, Princeton University Press, Princeton NJ, 2018.

Bjornerud, M. *Turning to Stone: Discovering the Subtle Wisdom of Rocks*. Flatiron Books, New York NY, 2024.

Khatwa, A. *The Whispers of Rock: Stories from the Earth*, The Bridge Street Press, London, 2025.

Macfarlane, R. *Underland – A Deep Time Journey*. Penguin Books, London, 2019.

Shepherd, N. *The Living Mountain: A Celebration of the Cairngorm Mountains of Scotland* (The Canons), Canongate Books, London, 2019 (originally published 1977).

Van Horn, G., Kimmerer, R. W. and Hausdoerffer, J. (eds). *Kinship: Belonging in a World of Relations*, Center for Humans and Nature, Libertyville IL, 2022.

ACKNOWLEDGEMENTS

Thank you to everyone at Quarto who helped me turn my vision into an actual book, especially John, who supported the idea right from the start, and my editor, Kat, who helped make this book the best it could be.

I would also like to thank the following. Stef, April and Rachel, for reading the drafts. Anna from Eòrna Pottery for reading and sharing her pottery knowledge. Emily for proofreading and letting me rummage through the rocks in the National Museums Scotland collections. Owen, Tery, my mum and dad for their never-ending support (and for tolerating my rock obsession).

Last but not least, thank you to the rocks that I have had the privilege to meet and get to know over the years. This book literally wouldn't have been possible without them.

The following specimens were kindly made available by National Museums Scotland as references for my illustrations: porphyry (Ancient Egyptian porphyry; G.1891.256.112), peridotite (forsterite; G.2000.16.1), anorthosite (Harris), tonalite (hornblende diorite; G.1893.24.121), granodiorite (G.2001.119.287.1), muscovite granite (68), syenite (G.2013.4.98), kakortokite (G.2013.4.93), lamprophyre (monchiquite; Loch Roag), carbonatite (G.2013.4.37), tachylyte (G.1996.43.2), snowflake obsidian (G.2005.4.62), pitchstone (G.1924.4.19), phonolite (G.1893.24.109), tuff (G.1885.34.79), kaolin (white clay; G.1858.212.160), mudstone (G.2008.35.16 and G.RC.UR.2986), siltstone (boulder clay; G.1998.43.272.1), loess (G.1912.6.11), marl with algae (G.1885.34–15), Rhynie chert (from the pile), tufa (G.1998.43.98.1), laterite (bole; G.RC.UR.2979), granulite (pyrope-almandine; G.2002.26.2253), eclogite (G.RC.UR.2748), quartzite (S.UR.56.15.M), lapis lazuli (lazurite; G.2000.40.1), jadeitite (jadeite gneiss; G.2011.37.2 and jadeite schist; G.2011.37.3), greisen (G.2013.38.2), unakite (epidotized granite; G.RC.UR.855), skarn (F.UR.14), pseudotachylyte (G.2014.36.9), impact breccia (suevite; G.2013.7.1), tektite (moldavite; G.1981.42.1 and australite; G.1862.54.2), coin in lava (G.1862.48.181 and G.2004.39.27).

ABOUT THE AUTHOR & ILLUSTRATOR

Vojta Hybl is a Czech-born illustrator, writer and educator living in Scotland. Originally trained as a geologist and physical geographer, he started using art to share the stories hidden in rocks and landscapes in 2021. Since then, his illustration practice has grown to include all parts of the natural world. Vojta combines his scientific background with art to create illustrations that reveal the interconnectedness of everything – and everyone – on Earth. Some of his clients include National Museums Scotland, The University of Edinburgh and Dynamic Earth, and his writing was published by the Scottish Mountaineering Press. *Rocks* is his first book.

↑
Photo by:
India Hunkin

INDEX

Quarto

First published in 2026 by Frances Lincoln,
an imprint of The Quarto Group.
One Triptych Place, London, SE1 9SH,
United Kingdom
T (0)20 7700 9000
www.Quarto.com

EEA Representation, WTS Tax d.o.o., Žanova ulica 3,
4000 Kranj, Slovenia
www.wts-tax.si

A catalogue record for this book is available from the British Library.

ISBN 978-1-83600-977-1
Ebook ISBN 978-1-83600-978-8

10 9 8 7 6 5 4 3 2

Design by Michelle Kliem

Publisher: Philip Cooper
Editorial Director: John Parton
Managing Editor: Laura Bulbeck
Editor: Katerina Menhennet
Senior Designer: Isabel Eeles
Senior Production Controller: Rohana Yusof

Printed in Huizhou, Guangdong, China TT042026